Exploring the U.S. on the Net

Cynthia G. Adams

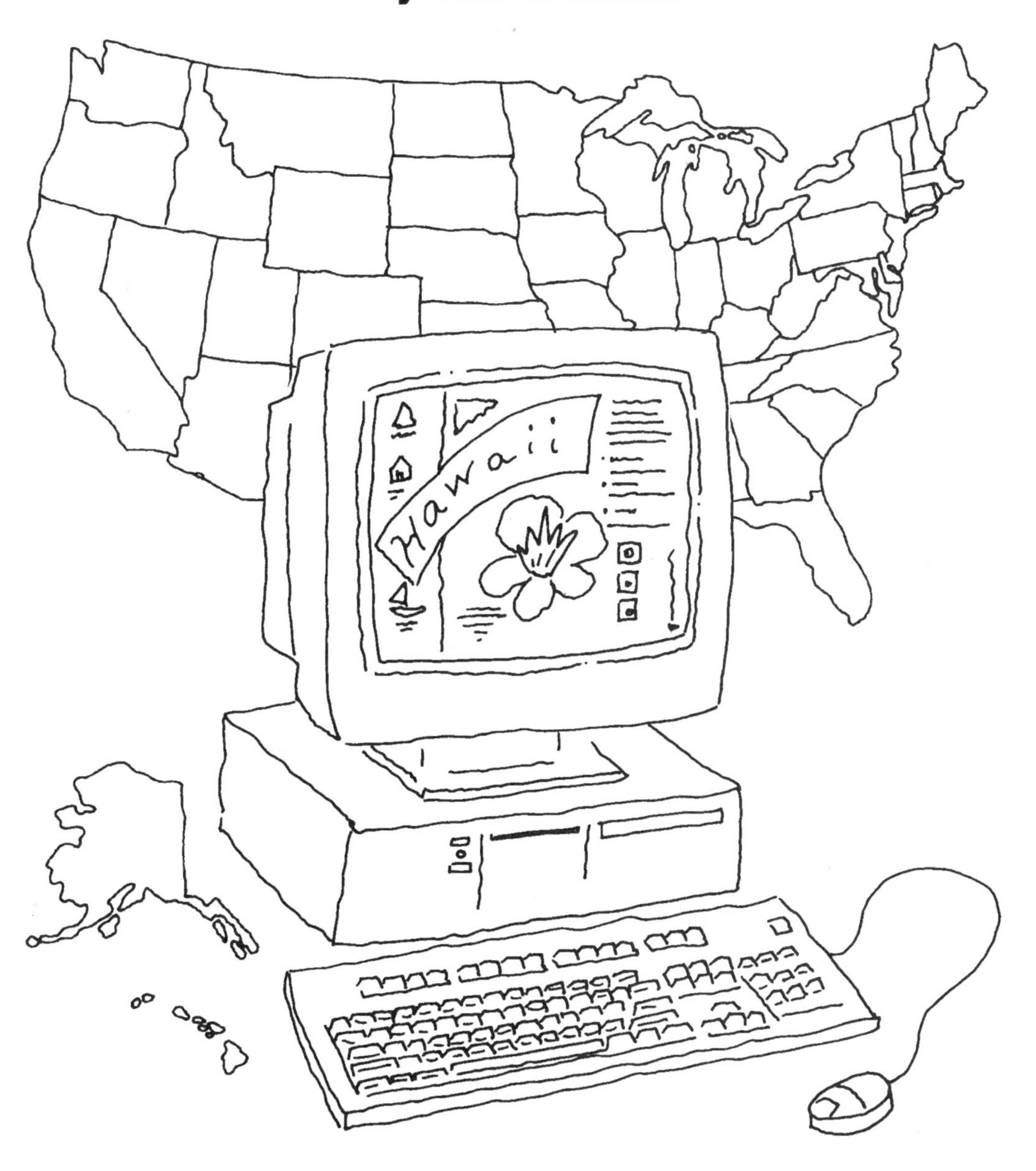

Good Year Books

Parsippany, New Jersey

Good Year Books

Good Year Books

are available for most basic curriculum subjects plus many enrichment areas. For more Good Year Books, contact your local bookseller or educational dealer. For a complete catalog with information about other Good Year Books, please write:

Good Year Books
An imprint of Pearson Learning
299 Jefferson Road
Parsippany, New Jersey 07054-0480
1-800-321-3106
www.pearsonlearning.com

Book design and illustration by Amy O'Brien Krupp.

Printed in the United States of America.

0-673-58638-3

4 5 6 7 8 9 - ML - 06 05 04 03 02 01 00

This Book is Printed on Recycled Paper

Contents

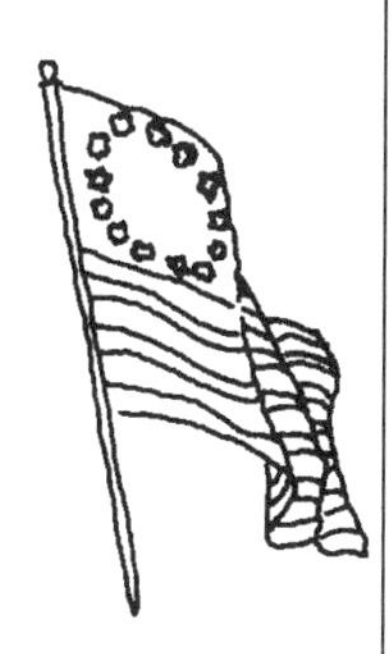

Introduction

E*xploring the U.S. on the Net* will introduce your students to the cities, landmarks, geography, culture, and history of the United States through use of the Internet. Much more than a directory, this book is designed for busy classroom teachers. Each page is divided into two reproducible task cards featuring usable Internet addresses and clear questions, activities, or project suggestions. The tasks encourage critical thinking and geography skills as well as learning. These practical ideas will complement your social studies and multicultural units and provide a valuable resource for individual enrichment or small group learning.

Before using the task cards in this book, familiarize yourself with the terms in the glossary. Once you understand these terms, you will be ready to use the Internet as a tool in your classroom.

Before You Begin

Because the Internet is always changing, it is a good idea for you to check each site before assigning it. You may find that an address has changed or a site has completely disappeared. In some cases, pages may have been moved or renamed. For this reason, many cards offer two similar addresses so that students' progress will not be slowed.

Check the sites for grade-level appropriateness and to determine a means of evaluating student progress. If you cannot access a Web site, you may try again at a later time, or you may be able to continue your search by shortening the address from the right. For instance, if **http://www.yahoo.com/Regional/U_S_States/Alabama/Education/K_12/Public_Schools/Middle_Schools/** does not work, try **http://www.yahoo.com/Regional/U_S_States/Alabama/Education/K_12/**.

After you have determined the best sites for study, your students will be able to search more efficiently if you organize *bookmarks* (or *favorites)* for each content area. Refer to your browser help screen or manual for exact directions on how to accomplish this. You may choose to distribute the list of General Sites on page 11 and require students to restrict their independent research to those sites.

Your experiences with the Internet in the classroom can be made more positive with careful planning and parent or peer help. They can interpret assignments and help if reading levels are too high. In most cases, all parents should submit a release agreement (p. 10) before students begin to use the Internet.

How to Use the Cards

The task cards can be duplicated, laminated, or cut apart, then placed in your computer center. You may prepare cards for a single state or section of the country. Students may use the cards during free time, as a structured assignment, or for enrichment according to individual interests and time constraints.

If you prefer, copy the cards and distribute them as consumable materials. Simply check off the specific activities or questions you wish your students to complete. If appropriate, indicate the date by which the assignment must be submitted. Depending on Internet accessibility, the cards may be used for homework assignments. Students may answer on the backs of cards or on separate sheets of paper. Direct students to sign the completed cards and return them to you for credit. Grade the cards according to classroom standards.

You may want to provide reproducible U.S. maps for students to use as they complete the activities on the cards. Several sites provide state maps that students can print and use for tracing routes or labeling landmarks.

As you introduce your students to the Internet using the cards on p. 9, be sure to emphasize that:

- every address begins with http://; and
- the site addresses must be entered accurately.

Students should begin by using the addresses on the task cards from the list of general sites. Students can locate additional information by clicking the mouse on the hypertext links within each of these sites.

The bulleted items on the cards are questions that can be answered with information that is readily available in the site. They are primarily reading comprehension questions or items that give practice in reading and following directions. Items with pencil icons are critical thinking questions or activity suggestions that will result in a tangible product. They are expanded activities for higher level thinking.

The boldfaced words on the cards match hypertext your students will click on to answer the questions. On the general information card of each state you will find "search words" to use for independent browsing. Advanced students may use the search words for extended research or if selected sites are no longer available. Use the bookmark or favorites feature of your browser to keep a record of the best Web sites on particular topics.

Using a Search Engine

Search engines provide listings and summaries of addresses on a given topic. You may select a broad category and refine it several times to locate your specific topic. Sometimes it is more efficient to type key words into the bar and push the Enter or Select keys. In most cases, the search engine will list many more sites than you could possibly visit. Take time to read the summaries before making a selection.

Page 7 teaches students how to use search engines. Page 8 provides an activity master for practice in using search engines.

Blocking Software

Software is available to block some of the inappropriate materials on the Internet, although none is fully effective because the Web changes so quickly. Be sure to check with your provider before downloading any of these products.

Cyber Patrol: Allows you to control access to categories of adult material and block words and phrases in chat rooms. Also has a clock that limits online time.
http://www.cyberpatrol.com

Cybersitter: Allows you to block foul language and keeps a list of sites a child has visited.
http://www.solidoak.com

Net Nanny: Comes with a list of blocked sites that is updated monthly. You can add to the list and block access to chat groups and certain words and phrases.
http://www.netnanny.com

How to Use a Search Engine

You may be asked to use a search engine for an additional study. Search engines provide listings and summaries of addresses on a given topic. You may select a broad category and refine it several times to locate your specific topic. To use a search engine, follow these steps.

1. Type and enter the address of the home page for the search engine of your choice. Here are addresses for some common search engines:

Yahooligans! (for kids)	**http://www.yahooligans.com/**
Yahoo!	**http://www.yahoo.com/**
Alta Vista	**http://altavista.digital.com/**
Excite	**http://www.excite.com/**
Lycos	**http://www.lycos.com/**

2. When the home page appears, type and enter key words that will lead you to the information you need.

For a difficult search you should start with a broad topic and gradually make it more specific. For instance, from the Yahoo! home page, you could select Alabama, Arts and Humanities, then History to find information about the Civil War, or simply type "Civil War, Alabama" into the bar and click the mouse on Select or Enter. Because the Civil War is a common subject, lots of information will be readily available.

Next, you must skim the listings to find the entry(ies) that most nearly fill your needs. Point and click your mouse for access to any Web site in the list.

Practice Using a Search Engine

Search words ______________________________
(teacher fills in)

Search engine(s) used ______________________________

Number of hits ______________________________

Web sites selected ______________________________

____

Five questions about the topic

1. ______________________________

2. ______________________________

3. ______________________________

4. ______________________________

5. ______________________________

Exchange your questions with a friend.
Answer them using information from the Web sites.

Answers

1. ______________________________

2. ______________________________

3. ______________________________

4. ______________________________

5. ______________________________

Cut along dashed line.

Introductory Card 1

Keypals and School Home Pages

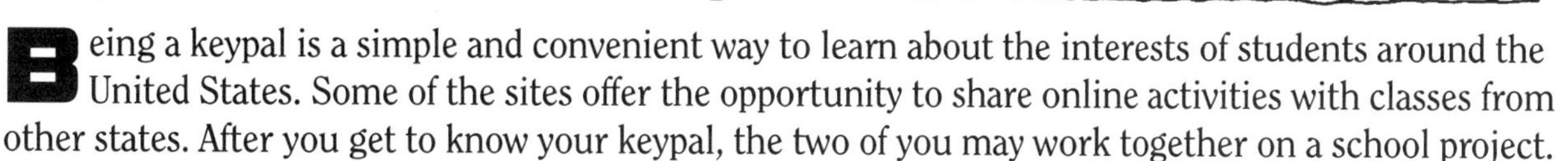

Being a keypal is a simple and convenient way to learn about the interests of students around the United States. Some of the sites offer the opportunity to share online activities with classes from other states. After you get to know your keypal, the two of you may work together on a school project.

You can use the following sites to select a student (or class) with whom you can exchange e-mail. You may select and visit school home pages around the country from these sites. Click on classes for closer a look. Most of them ask you to indicate your name, age, and interests.

Web sites: **http://www.geocities.com/SunsetStrip/6460/penpals2.html**

http://www.oink.demon.co.uk/penpals.htm

http://www.yahooligans.com/School_Bell/Elementary_Schools/

Introductory Card 2

United States and Major Cities

Web site: **http://www.city.net**

You may search any state or city in the United States from this site.

1. From the home page select **North America,** then **United States.**
2. Next you will see a map of all the states. Make your selection.
3. When you arrive at the home page of the state you will find an index that includes maps, travel/tourism information, city guides, weather, recreation and entertainment suggestions, and photo galleries.

Search the state where you live, a city you have visited, or choose any city you would like to know more about. What would a tourist enjoy about this location? Find an address for one hotel, restaurant, and a museum or monument in the city. Describe the city's geographic location within the United States and indicate the current weather conditions.

Parental Release Agreement

Dear Parent:

We are pleased to announce that Internet access is now available to your child in our classroom. We are excited to be able to provide our students with this valuable learning resource.

Your child will be given assignments to be completed using the Internet. Because the Internet may include some inappropriate material, you may be assured that we will carefully supervise and monitor your child's use.

Students will be given clear guidelines for using the Internet and specific assignments to follow. All students understand that:

- accessing unacceptable materials is strictly forbidden;
- using inappropriate language is strictly forbidden;
- revealing their addresses and phone numbers is forbidden; and
- e-mail is not private.

Please read the above agreement with your child, sign, and return to the teacher by ______________ .
(date)

✂ Cut along dashed line.

I have read the terms and conditions for using the Internet.
I realize that a violation will end my access privileges.

Student ________________________________ Date ____________________

I understand that it is impossible for the school to restrict controversial materials on the Internet. I have discussed appropriate use of the Internet with my child and give my permission for him or her to access the Internet at school.

Parent or Guardian __________________________ Date ____________________

General Sites

Here are some general Web sites that may be useful with your study. You may want to include them as bookmarks.

http://www.yahooligans.com/School_Bell/Elementary_Schools/
http://www.yahooligans.com/School_Bell/Middle_Schools/
These present home pages of students from various states.

http://web66.coled.umn.edu/schools.html
Use the clickable map to visit home pages from selected states.

http://www.yahooligans.com/School_Bell/Field_Trips/
Share field trip experiences with schools around the country.

http://www.yahooligans.com/School_Bell/Homework_Answers/Geography/
See a variety of maps from each state from this site.

http://femi.jhuapl.edu/states/states.html
This site has a clickable directory of states that leads to a variety of map links.

http://www.classroom.net/classweb
Create your own school page and post it free of charge from this site. Search the classroom archives for a school link near you.

http://www.ipl.org/youth/stateknow/
Choose any state and quickly get basic information. Also includes tables for size and population rankings as well as games based on state flags and capitals.

http://info.er.usgs.gov
This site will help you understand mapmaking. Includes satellite images of different parts of the world.

http://www.lib.utexas.edu/Libs/PCL/Map_collection/histus.html
Use this site for historical maps of the United States.

http://quest.arc.nasa.gov/
This is NASA's effort to involve classrooms with use of the Internet.

http://www.cyberkids.com/
Play games or share original compositions, art, and writing projects with students around the country at this site focusing on creativity.

Extension Ideas

Fine Arts Projects

- Design a travel brochure and poster describing one state.
- Build a model of a landmark in a state.
- Draw a time line depicting a hundred years in the history of one state.
- Think of at least three ways one state has contributed to the culture of the United States (or the region).
- Design a set of six postage stamps commemorating natural landmarks or famous people from a specific state.
- Design a postcard from a trip to a state. Write about what you've seen and experienced.

Writing Activities

- Pretend you are the governor of one state. Write two laws (or reforms) that would improve life for the people.
- Choose a famous landmark in a state. Describe it in detail. How is the landmark significant to the local people?
- How does tourism affect the economy of a specific state?
- Write atlas entries for three different states.
- Write a diary of a week's vacation in one state.

Social Studies Projects

- Role-play a meeting between two famous people from one state. What is their significance to the past or the present? Pose a current problem for them to resolve.
- Compare the characteristics of two states in the same region of the country.
- Compare the cultures of people living on the East and West Coasts.
- Pretend you have just returned from a state. What would you most want to tell someone who is planning a visit there?
- Invent two products made from one state's natural resources.

Mapping Activities

- Trace the most direct path from your home to a specific state. Which direction(s) would you go and what states would you pass over?
- Draw a map of one state. Label five major cities or regions.
- Make a topographical or climatological map of one state. Explain how topography and climate influenced population movement and the settlement patterns of the state.

Task Cards

Alabama

General Information

Web site: http://isl-garnet.uah.edu/ALABAMA/alabama.html

- Use the chart for basic information about Alabama. On what date was Alabama admitted to the Union? What is the state bird? the state flower?
- Select **Alabama.** Use the clickable state map to select and explore the county of your choice. Make a chart showing the area, population, and county seat of six different counties. Indicate landmarks and major cities.

Web site: http://alaweb.asc.edu/

- Click on **All About** and **Quick Facts.** What are the three largest cities?
- What are the chief agricultural and manufactured products?
- How do service industries affect the state's economy?
- Click on **Frequently Asked Questions.** How could you contact Alabama's governor or state senator?

Alabama

Cities

Web sites: http://www.ci.bham.al.us/
http://www.bham.net/bcri/index.html

- Read about the history of **Birmingham,** Alabama's largest city.
- Make a list of the attractions available to tourists.
- ✎ Select the **Sports Hall of Fame Museum.** What famous Alabamans are immortalized there? Categorize athletes by sport and do research to indicate teams or years of professional play.
- Explain how the Civil Rights movement affected Birmingham and Montgomery in the 1950s and 1960s. What special programs are available to people who visit the **Civil Rights Institute?**

Web site: http://www.mongomery-al.com

- Click on **Our City,** then **City Maps.** What two interstate highways come together at Montgomery?
- Use the **Newcomer's Guide** to determine what industries are common in the area. How many colleges and universities are within an easy commute of the city?

Cut along dashed line.

Cut along dashed line.

Alabama

Tourist Attractions

Web site: http://www.state.al.us/ala_tours/tours.html

- Click on **Hot Spots** and take a tour of the state capitol.
- Go to **Annual Attractions.** Choose three you would like to visit.
- ✎ Make a brochure describing one attraction. Indicate admission fees and hours of operation.
- Select **Bellingrath Gardens and Home** and click on **What's in the Gardens.** Read about its **History.** Take a virtual tour using the clickable map.
- ✎ Is there a public garden or conservatory in your city or town? How does it compare to Bellingrath? Make a list of flora.
- ✎ Compare Bellingrath and Middleton Gardens in South Carolina.
- Return to the home page and click on **State Parks.**
- Visit the site of one park and explain why you think it would be a good place for a vacation. Locate the park on a map and determine the nearest city.

Web site: http://www.touralabama.org/

- Select **What to See and Do,** then **Attractions by Category.** Use the menu to select an item that interests you or your family. From many of these sites you can learn about other **area attractions.**

Alabama

Geography

Web site: http://www.state.al.us/general/general2.html

- What is the **soil** composition in Alabama?
- ✎ Draw a map showing the soil type in each **region** of the state. Indicate which region is best for agriculture, illustrating the products with icons.
- What states are to the north, south, east, and west of Alabama?
- Go to **climate.** What are the average temperatures during the summer and winter seasons? What is the average annual precipitation?
- ✎ Send an e-mail and receive a student information packet.

Alaska

General Information

Web site: http://www.alaskasbest.com/facts.htm

- This site has an extensive list of **Fun Facts about Alaska** and will serve as an excellent introduction to the state. What is the state bird? flower? gem? mineral? fish? sport?
- Select **Great Outdoors.** What is the current range of temperatures in Alaska? Choose individual cities for more specific weather information.
- ✎ Write a one-week travel diary for a virtual Alaskan adventure. Check **Outdoor Tips** for safety considerations in the wilderness.

Web site: http://www.corecom.net/ficus/alaska.html

- Click on **Frequently Asked Questions.** When and where was the heaviest recorded snowfall in Alaska? What are ice worms?
- ✎ If you have a question about Alaska that you need answered, you may send e-mail from this site.

Web site: http://www.alaskasbest.com/

- At this site sponsored by *Alaska's Best Magazine,* you can find **weather information** about the state.
- ✎ Make a graph showing temperature information for any five cities.

Search words: Kenai Peninsula, gold rush, Yukon, Aleuts, Eskimos, Glacier Bay, Mount McKinley, Denali National Park, Skagway, Iditarod, Unimak Island

Alaska

Cities

Web sites: http://juneauempire.com/Guide/faqg.htm

- Read the **facts about Alaska's capital city.** How does it compare to the capital of your state? Make a chart comparing the population and total square miles of five different states.
- Use the **Links** to go to **Juneau's home page.** Who is the governor of Alaska? the lieutenant governor?

Web sites: http://usacitylink.com//juneau/default.html
http://www.onroute.com/destinations/juneau.html

- Plan a virtual trip to Juneau. Make a list of what you can see and do and decide on a **place to stay.** How would you describe the "true Alaskan experience"?

Web site: http://www.ci.anchorage.ak.us/

- List six reasons why Anchorage is a great place to live.
- Click on **History of Anchorage** to view a time line. Write a narrative about fifty years in Alaskan history.

Cut along dashed line.

Cut along dashed line.

Alaska

Tourist Attractions

Web site: http://www.juneau.com/guide/photos/

- Enjoy this set of photographs from the **Department of Fish and Game.**
- ✎ Choose one of the photos to print. Search the Internet for additional information and write a brief report. Combine all the reports in a class book of *Alaskan Fish and Game.*
- Select **Alaska Gold Rush.** What general statements can you make about the gold rush from viewing these photos? How did the miners live? What did they do for entertainment? What activities are popular with tourists?

Web site: http://www.alaskanet.com/Tourism/index.html

- This is the site to visit before planning an Alaskan vacation. Search the extensive database to schedule ground or air travel. List the five national parks in Alaska. What accommodations are available for tourists? Which park would you choose to visit? Explain.
- ✎ Choose one park and search the Internet for more information. What fish and game can be seen there? What is the geography of the park?

Web site: http://www.skagway.org/

- Visit this reproduction of a frontier town during the gold rush. Click on the icons to read about the city's history, event calendar, and accommodations.

Alaska

Geography

Web site: http://www.juneau.com/guide/maps/

- Click on **Alaska State Map.** How many miles would you travel from Nome, Alaska, across the Bering Sea to Russia? What country borders Alaska to the east?
- ✎ Click on **Alaskan Volcanoes.** Choose one classification from the map key. Select ten different volcanoes. Create a chart showing the years they erupted.

Web site: http://www.city.net/countries/united_states/alaska/

- ✎ Make a graph of high and low temperatures in Anchorage, Fairbanks, Juneau, and Ketchikan.

Web site: http://www.ptialaska.net/~pongo/northern/

- From this site you can visit communities that experience the **Northern Lights.**

Web site: http://www.alaskagold.com/valez/index.html

- Explain the **glaciation** at Prince William Sound. Why did the glaciers form? How have they shaped the land?
- Use the **Links** to read about the most recent Iditarod.

Arizona

General Information

Web site: http://dizzy.library.arizona.edu/images/library-images.html

- This is a wonderful site with photos and information on the folk art and religious traditions that make Arizona unique. Click on **Mission Churches of the Sonoran Desert** to read about the history of the area and the Tohono O'odham people. Look at the mission photos and describe the building style.

Web site: http://www.arizonaguide.com/

- Click on **State Flag** and **History.** Read the information.
- ✎ Draw an Arizona flag. Explain the symbolism of the design.
- People from what European country first settled the Arizona territory? In what year did Arizona become a state? What minerals attracted early settlers to the state?
- Select **State Info** and read about the American Indians who make Arizona their home. What groups live there? Name three of the prehistoric dwellings that are now national monuments.

Search words:
Grand Canyon, Anasazi Indians, Taliesin West, Tempe, Scottsdale, Pueblo Indians

Arizona

Cities

Web site: http://www.city.net/countries/united_states/arizona/phoenix/

- Click on **Sightseeing.** What is the address of the Visitor Information Center? Where would you go to see exhibits of American Indian crafts? pioneer homes? the world's largest desert plant collection?
- ✎ Search the Internet for information about Taliesin West. What is it, and who was its famous architect?

Web site: http://www.virtualperfection.com/Tucson.html

- From this site you can take a virtual tour of Tucson, its resorts, and its restaurants. How does the blending of Mexican, Spanish, and American Indian cultures affect the city's style?
- Use the **maps** to find the locations of Saguaro National Monument, Coronado National Forest, and the Tohono O'odam American Indian Reservation.
- ✎ List examples of the cultural influences of the Spanish and American Indians. These may be street or city names, food preferences, craftspeople, or building styles.

Cut along dashed line.

Arizona

Tourist Attractions

Web sites: http://www.kaibab.org/
http://www.azstarnet.com/grandcanyonriver/GCrt.html

- Take a virtual tour of the Grand Canyon area, or a trip down the Colorado River at this award-winning site.
- ✎ Use a map and chart the route to the canyon from the south (Flagstaff), west (Williams), or north (Kanab, Utah).

Web site: http://www.thecanyon.com/nps/backcountry/index.htm

- Click on **Trip Planner.** What are the fees for lodging on the South Rim? North Rim? Describe a mule ride from the South Rim. Would you take a mule ride? What number should you call for reservations?

Web site: http://www.arizonaguide.com

- The links at this site will guide you to information about coming events in each area of the state.
- ✎ Check the **Weather Links,** then choose a location to visit. Think about what items you would need to pack for an Arizona vacation. Make a list of items, then trade your list with a friend and make additional suggestions.

Arizona

Geography

Web site: http://www.azinfo.com/map1.ipg

- Use this map to trace a route from Tucson to Phoenix. What highways would you use? How many miles is it between the two cities?

Web site: http://www.pagehost.com.flagstaff/index.html

- ✎ Make a line graph showing the high and low average daily temperatures in Flagstaff, Arizona.

Web site: http://www.execpc.com/~drogge/atlas.html

- Use this Arizona picture atlas to visit famous landmarks. List the animals and plants indigenous to the area.
- ✎ Search the Internet for additional information on the saguaro cactus.

Web site: http://www.arizonaguide.com/

- Select **Geography.** What is Arizona's geographic location? What states occupy its northern, western, and eastern borders?
- ✎ Design a state map showing three distinct regions of Arizona, and include information about the flora and fauna of each. Include a map key.

Arkansas

General Information

Web site: http://www.arkdirect.com/

- This is the search engine for the state of Arkansas. From here you can access information about education, government, sports, and tourism. You may also type in a key word to do a more specific search.
- Access the home page of the state government. Who is the governor? lieutenant governor?
- Select **Arkansas Heritage.** What kinds of things does the Heritage Commission preserve and encourage around the state?
- ✎ How does our heritage affect the way we feel and respond to nature and art? Cite examples from your own experience to support your answer.

Web site: http://ourworld.compuserve.com/homepages/mikenewman/indexark.htm

- Use **Arkansas Facts** to discover which is the largest city. What is the population of the state? When was Arkansas admitted to the Union?

Web site: http://www.whitehouse.gov/

- Read about the life of William Jefferson Clinton, the first U.S. president from Arkansas. What were his most important accomplishments as governor?

Search words:
Ozark Mountains, Fort Smith, Razorbacks, Buffalo River

Cut along dashed line.

Arkansas

Cities

Web site: http://www.arkdirect.com/

- Click on **Travel and Tourism,** then **Cities,** then **HometownNet.** From this page you can access home pages of many cities and towns. Look at five different sites. In which city would you prefer to make your home? Give five reasons to support your answer.

Web site: http://www.littlerock.com/lrcvb/

- Select **Area Attractions.** What kinds of (indoor) entertainment are popular with the people of Little Rock?

Web site: http://www.weather.com/weather/us/cities/AR_Fayetteville.html

- What is the five-day weather forecast for Fayetteville? Use the search bar to research conditions in at least three other Arkansas cities. Use this information to make general statements about current Arkansas weather.

Cut along dashed line.

Arkansas

Tourist Attractions

Web site: http://www.arkdirect.com/

- Select **Travel and Tourism** and visit the **Ozark Mountain Region.** Follow the **Tour Guide,** and print a map of the **driving tour** you prefer. Read the information about the tour you printed and then circle five locations on your map.
- Click on **Site Map** to see the five-county region. Circle the area of your tour. Which interstate highway runs east to west? What road(s) would you follow to drive north to south?

Web site: http://www.hotsprings.org/index.html

- Click on **Tourism** to discover the appeal of Hot Springs, Arkansas. Check the **Calendar** for events that are happening this month. Select **Attractions** to read what is available for families.
- ✎ Write a persuasive paragraph telling why Hot Springs is or isn't a good family vacation destination. Explain three reasons for your feelings.

Web site: http://www.yournet.com/

- ✎ Use this site for a listing of all **state parks.** Read the charges for renting a cabin in one of the state parks and estimate the cost of a one-week stay (including a budget for food).

Geography

Web site: http://www.ozarkmtns.com/

- What lakes and rivers are located in the Ozark Mountain Region? Which lake was formed by a humanmade dam across the North Fork River?
- Select the **Buffalo River** and look at the map of the river. Select the **Floater's Guide** to read about the characteristics of the river. Where does the river originate? Do you think it would be easy to canoe on this river? What would be the best time of year to backpack and fish the Buffalo River?
- ✎ Create a map of Arkansas. Draw the Ozark Mountains and the Buffalo River.

Web site: http://www.littlerock.com/lrcvb/facts.html

- ✎ Check the chart of **driving times** between several major cities and Little Rock, Arkansas. Determine the state and locate each city on a U.S. map. Draw lines to connect the cities to Little Rock. Which city is nearest? farthest? Which city is closest to you? Connect your hometown to Little Rock and write the approximate driving time.

California

General Information

Web site: http://www.ca.gov/s/

- Click on **Government.** Who is the governor? What are his or her duties and responsibilities? List four ways the citizens of California can contact the governor.
- ✎ Write two good questions that a voter might ask the governor.
- Click on **Natural Resources.** What are Californians doing to protect their environment? List ten suggestions for "Going Green."

Web site: http://www.focusoc.com/state/index.html

- Read **State Rankings.** Choose one item from the **first rankings** that has a negative effect on California's citizens. Consider the causes of the situation and think of three ways to correct it.
- Read the information in **markets.** How much money does the California economy generate in goods and services each year?
- ✎ Make a bar graph showing the dollar amounts attributed to each of the major industries.

Search words:
Mission San Juan Capistrano, Hearst Castle, redwood forest, Napa Valley, Sausalito, Alcatraz, Hollywood, Golden Gate Bridge, Big Sur

Cut along dashed line.

California

Cities

Web sites: http://la.yahoo.com/

- At the Yahoo! search engine dedicated to Los Angeles, click on **Maps and Views** to see live views and photos of the city and **Points of Interest** for local maps.

Web sites: http://sfbay.yahoo.com/

- The first site is the Yahoo! search engine dedicated to the San Francisco Bay area. Use **Maps and Views** for live views and photos.
- Print the **Bay City Guide Map.** Name five north-south streets and five east-west streets. Describe the location of Ghirardelli Square and the Cannery.

Web site: http://www.santabarbaraca.com/

- Explore Santa Barbara from this site. Click on **Information,** then **History** to read about this unique city. What European explorer first claimed the land for Spain?
- ✎ What is an adobe? Search the Internet for information about this form of architecture. Draw an example.

Cut along dashed line.

California

Tourist Attractions

Web site: http://www2.disney.com/Disneyland/index.html?GL=H

- This is the official home page of **Disneyland.** Why do you think Walt Disney chose this location for his first amusement park?
- ✎ If you have been to Disneyland or Disney World, tell the class about your experience. Compare the facilities at Disneyland and Disney World using information available on the Internet.

Web site: http://www.donaldlaird.com/landmarks/

- Click on a county name and see photos of the specific historical landmarks. Select a landmark, then directions, and read the address. Draw a California map and label twenty landmarks and cities.

Web site: http://www.tsoft.com/~cmi/

- From this site you can access information about the people and history of **California's Spanish Missions.** You can take a virtual tour of a typical mission. The photos can be viewed in order, or you may take a quick jump to the photo of your choice.
- ✎ Use the **Quick Jump** menu and the map to create your own map of a typical mission. Label your diagram. You may include a native village. What was the original purpose of the missions?

California

Geography

Web site: http://www.lalc.k12.ca.us/laep.smart/river/tour/index.html

- This site will lead you on a descriptive tour of the **Los Angeles River,** from the San Fernando Valley to Long Beach. Twelve sites present photos and information about plants, animals, architecture, and history.

Web site: http://www.scecdc.scec.org/eqsocal.html

- You may search this site for a map of the faultlines of Southern California. Read information about individual earthquakes by clicking on the location on the map.
- Click on **Putting Down Roots in Earthquake Country** to learn about safety precautions. What is an aftershock? a foreshock?
- ✎ Most people accept the risk of earthquakes because they enjoy living in Southern California. In your opinion, do the benefits outweigh the risks? Explain.

Colorado

General Information

Web site: http://www.mii.com/colorado/index.html

- Use this site to access information about the cities, recreation, weather, and businesses of Colorado. Read the facts. What is meant by the motto "Nothing Without Providence"?
- ✎ Search the Internet for pictures of Colorado's state flower, bird, and tree.
- Click on **Weather.** What are current weather conditions in the capital city?
- Click on **Government,** then **Citizen's Guide to Colorado.** Where would you phone to get help with child care?

Web site: http://www.geocities.com/Yosemite/3913/Imgcam.htm

- A great site to view lots of photos of Colorado resorts and cities. Click on the **Denver Eye 9 Cam Network** to see weather shots of five different locations.

Search words:
Aspen, Vail, Telluride, Fraser River Valley, Rocky Mountains, Pikes Peak, U.S. Air Force Academy

Cut along dashed line.

Colorado

Cities

Web site: http://www.infodenver.denver.co.us/

- Check the **News** from Denver. How did the city get its name? What is the symbolism of the Denver flag?
- Click on **Local Government.** In what year did Denver become the capital city of Colorado? What are the responsibilities of the city council? How many people currently live in the city of Denver? in the county?

Web site: http://bcn.boulder.co.us/boulder/

- Read the **History of Boulder.** How did gold miners contribute to the settlement of the area?

Web site: http://www.colorado-springs.com/

- When was gold discovered at Cripple Creek? What is the estimated population?
- ✎ Explain why the climate and snowfall of Colorado Springs are desirable compared to conditions in Denver. What are the Chinook winds? What is Colorado Springs meteorological classification?
- Click on **Community Facts,** then **Business and Military.** Explain the military presence in Colorado Springs. What is unique about Cheyenne Mountain Air Force Base? What products are manufactured locally?

Cut along dashed line.

Colorado

Tourist Attractions

Web site: http://www.toski.com/

- This page has connections to all the popular ski and golf resorts in Colorado. Read about **Aspen, Telluride,** and **Vail.** Which of the three resorts is the most economical? Which has the best programs for children? How far is each resort from Denver?
- ✎ Draw a map of Colorado. Label three ski resorts and three golf courses.

Web site: http://www.vtinet.com/14ernet/

- Click on **Area Information.**
- The **Fourteeners** are a group of fifty-four peaks that rise more than 14,000 feet above sea level.
- ✎ Make a bar graph showing the elevations of any ten peaks.
- Go to **Land of Diversity** and select **Photo Album.** Choose one community. Explain why it would be a good tourist destination.

Web site: http://www.softronics.com/peak_cam.html

- Click on this **live cam** view of Pikes Peak. Who first discovered the mountain? Who was the first to reach the top? How can visitors reach the summit?

Colorado

Geography

Web site: http://bcn.boulder.co.us/boulder/bmp/history.html

- What **trees and shrubs** are native to the Rocky Mountain region? What wildlife species live in the Boulder Mountain Parks? What advice should hikers follow if they confront a **black bear** or **mountain lion?**
- ✎ Make a trail guidebook of eight to ten wildflowers native to the Boulder Mountain Parks.

Web site: http://www.csn.net/~arthurvb/COLORA.GIF

- This site has an easy-to-read road map of Colorado. What routes would you travel from Denver to Mesa Verde National Park? Approximate the number of miles from Denver to Colorado Springs. What direction would you travel on each trip?
- ✎ Print the map and trace the routes.

Web site: http://www.state.co.us/

- Seven different sites contain information about the communities in each region of Colorado. Locate travel directions and addresses for tourism offices.
- ✎ Read the **Fun Facts about Colorado.** Make a chart or booklet that illustrates ten facts.
- ✎ Choose one of the scenic byways to contact for travel brochures. You may use the mail or 800 telephone number.

Connecticut

General Information

Web site: http://www.state.ct.us/

- Use the **Frequently Asked Questions** to locate the e-mail addresses of the governor and attorney general. Use the link to visit the **Governor's Home Page.** How does the governor's office deal with issues of safety, welfare, and economic growth?

Web sites: http://www.ipl.org/youth/stateknow/ct1.html

- Use these sites to identify Connecticut's state bird, flower, and tree. Read the history of the state and explain why it is nicknamed the Constitution State.
- Click on **More Facts About Connecticut.** Make a list of famous people from Connecticut and describe the accomplishments of each one.
- ✎ Write three headlines that might have appeared in the *Hartford Courant* (the oldest U.S. newspaper still being published) describing events of the Revolutionary War that involved Connecticut.

Search words:
New England,
Litchfield Hills,
American Revolution

Connecticut

Cities

Web site: http://www.state.ct.us/MUNIC/HARTFORD/hartford.htm

- This is the home page for Connecticut's capital city. What is the size and population of Hartford? Do research to find out why Hartford has been nicknamed the Insurance City.
- Select **Bridgeport,** Connecticut's largest city. Locate it on a map and explain why its location helps support the factories located there. Read the information about the city's seal and motto.
- ✎ Compare life in Bridgeport with Stratford or Norwich.

Web site: http://utrcwww.utc.com/UTRC/General/Maps/areamap.gif

- Print the map of Hartford. What highways would you travel to move north-south and east-west through the city? Highlight each route.

Web site: http://ci.westport.ct.us/

- Click on **History.** How did the Revolutionary and Civil Wars affect the city of Westport? How did the city change in 1910?
- Take the photo tour of Westport.
- ✎ Search the Internet to learn what a doughboy is. What is a minuteman?

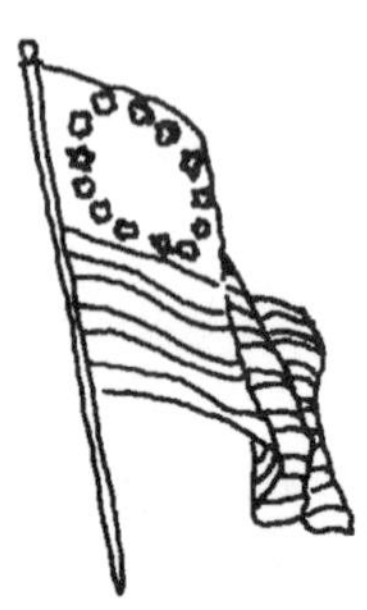

Cut along dashed line.

Cut along dashed line.

Connecticut

Tourist Attractions

Web sites: http://www.state.ct.us/tourism.htm

- Select **Vacation Guide** for an extensive group of links about Connecticut's tourist attractions. Make a chart showing what's available in each category.

Web site: http://www.litchfieldhills.org/

- How far is this region from New York City? from Boston?
- Read to find out **things to do and see** in the **Litchfield Hills region of Connecticut.** Search the site to find activities available for vacationers.
- ✎ Would your family enjoy a one-week vacation in the Litchfield Hills? List one thing you could do together each day of the week.

Web site: http://www.housatonic.org/

- Read about the composer **Charles Ives.** In which city is his birthplace? What techniques did he use to create a uniquely American style of classical music?
- Read about the **Historic Sites** and **Points of Interest** in the area. What activities probably occurred in these cities in the 1700s? Support your ideas with specific examples.

Connecticut

Geography

Web site: http://www.connecticut.com/overview/

- Go to **Geography.** What are Connecticut's three major rivers? Mark them on a map. Color the Litchfield Hills and the beaches along the shores of Long Island Sound.

Web site: http://www.connecticut.com/tourism/info/climate.html

- Describe the general **climate** of Connecticut. How do the northwestern hills affect the temperature in the central valley?
- ✎ Use the table to make a bar graph showing high and low temperatures for one year in Hartford.

Web site: http://dep.state.ct.us/Waste/recyfact.htm

- Explain the mission statement of the Department of Environmental Protection in your own words. Go to the **Table of Contents,** then select the **Recycling Fact Sheet.** What nine items can be recycled? What happens to the remaining trash before it is taken to landfills? What else is the state doing to encourage people to save space in landfills?
- ✎ Make a list of ten overpackaged items that you and your family use. How can people use lawn cuttings and food trimmings on their own property?

Delaware

General Information

Web site: http://www.state.de.us/govern/agencies/dedo/dsdc/dsdc.htm

- Click on **Geography,** then **1990 Delaware Census Geography.** Access the **Population by Age and Race** tables from the **State Division.** Compute the percentages of Caucasians, African Americans, and Hispanic Americans within the population of the state.

Web site: http://www.state.de.us/facts/history/delfact.htm

- Use this site to access basic information about Delaware, including the state bird, tree, flower, fish, mineral, bug, and beverage.
- What do symbols on the state seal signify?

Web site: http://www.deldirect.com/

- This site lists calendars of business and tourism events in the state. Check what's going on this month. What cities would you visit? To what events would you go?
- ✎ Pretend that you are an executive of one of the companies that has its headquarters in Wilmington. Plan a three-day itinerary for a business trip to Delaware.

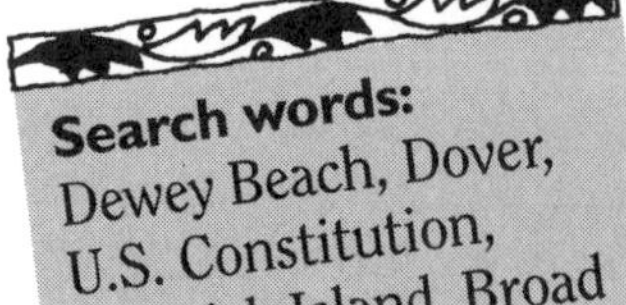

Search words:
Dewey Beach, Dover, U.S. Constitution, Fenwick Island, Broad Dyke Marsh, Kalmar Nyckel

Delaware

Cities

Web site: http:/www.udel.edu/Newark/DE/usa.htm

- This site enables you to visit the city of Newark. What kinds of information are available to new residents in the **Frequently Asked Questions?** List five things a landlord must do to maintain rental property. What is an alderman?

Web site: http://www.ci.wilmington.de.us/

- Click on **General Info,** then **History.** How did the Civil War affect the economy of Wilmington? How did the established industries aid the war effort?
- Go to **Facts and Figures,** then **City Profile.** What cultural and recreational opportunities are available in the city?

Web site: http://www.state.de.us/cities/dover.htm

- Locate each of the historic homes and museums on the street map. Choose two sites and write directions for walking from one to the other.
- ✎ Locate the Golden Fleece Tavern. What happened at that site on December 7, 1787? Do research to find the name of one man from Delaware who signed the Constitution.

Cut along dashed line.

Cut along dashed line.

Delaware

Tourist Attractions

Web site: http://www.dnrec.state.de.us/parks/dsp1st.htm

- Use the **Locator Map** to find out where most of the state parks are located. Which park is nearest Dover? Which park is on the Maryland border? Which ones are along the Atlantic coast?
- ✎ Draw a map of Delaware and label the fourteen parks.
- Read about the **Greenway Program.** What recreational activities do people enjoy on the Greenways?

Web site: http://www.deweybeach.com/

- Click on **Sights and Sounds** to stroll the beach or explore the town. Why do you think people on the East Coast consider this a great weekend getaway? Where would you choose to stay and eat if you were visiting Dewey Beach?

Web site: http://www.atbeach.com/

- Click on **Delaware,** then **Weather.** What are the current conditions at Bethany Beach? at Rehoboth Beach? Given the weather, what will people be doing at the beach today?

Delaware

Geography

Web site: http://www.udel.edu/delaware/map.html

- Use the interactive map to learn more about the cities and resources of Delaware. Choose a city to study in depth.
- ✎ Design a magazine advertisement for a feature of your city.

Web site: http://www.state.de.us/govern/governor/accomps/enviro.htm

- Read about what the government is doing to protect Delaware's beaches and wetlands. What has been done to protect endangered fish and birds?
- ✎ What are wetlands? Locate and identify wetlands in three other states.

Web site: http://city.net/img/magellan/mgmapsoftheworld/delawa.gif

- If you follow Route 113 from Dover, Delaware, which direction would you travel to reach the state of Maryland? Name three cities you would pass. Name a city on the Delaware-Pennsylvania border. What body of water lies between Delaware and New Jersey?

Cut along dashed line.

Florida

General Information

Web site: http://www.ego.net/us/fl/index.htm

- What is Florida's nickname? When did Florida become a state? Use the clickable map for quick connections to Florida's major cities. What city is farthest north? How would you reach Key West?

Web site: http://www.floridajuice.com/

- This site of the Florida Department of Citrus explains the steps to produce orange juice from tree to glass. You will find photos to enlarge and text with links to other citrus information.

Web site: http://www.erols.com/lthouse/home.htm

- Check the many lighthouses in Florida on the **site map.** You may take the tour of **Cape Florida Light** for an in-depth look at the parts of a lighthouse.
- ✎ Make and label a diagram of a lighthouse using information from the tour.
- ✎ Explain the purpose of and need for lighthouses along Florida's coast.

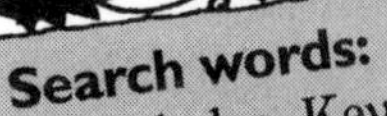

Search words: Everglades, Key West, Palm Beach, Sanibel and Captiva Islands, manatee

Florida

Cities

Web site: http://www.orlando-guide.com/maps.html

- Use the city map of Orlando to discover which major highways cross in the center of the city. What direction would you travel from the airport to Disney World?

Web sites: http://www.florida.com/daytona/day_alm.htm
http://florida.com/cities.htm

- Click on **Almanac.** How many miles of coast does Daytona Beach include? In what county is Daytona Beach? What is Daytona's connection to the development of automobile engines?
- ✎ Print a map of Florida. Draw a line to connect Daytona Beach to the five attractions at the end of this site. Indicate the mileage.

Web sites: http://www.florida.com/st_augustine/aug_alm.htm
http://florida.com/cities.htm

- List five reasons why St. Augustine is a distinctive American city.
- ✎ Search the Internet to discover the identity of Ponce de Leon. What is the Mission of Nombre de Dios?

Cut along dashed line.

Florida

Tourist Attractions

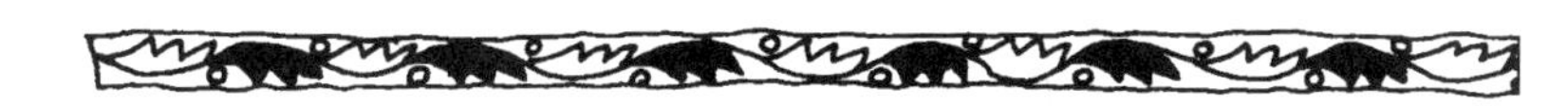

Web site: http://www.goflorida.com/

- This huge collection of links has tourist information, weather forecasts, and customized driving instructions and maps.
- ✎ Select one location from the Florida cities listed at the bottom of this site. Collect information about the city and explain to the class why it is a popular tourist destination.
- ✎ Draw a Florida map. Label tourist destinations in each area.

Web sites: http://www.spaceportusa.com/
http://www.ksc.nasa.gov/ksc.html

- Click on **Today at Kennedy Space Center** for information about curent space programs, then use **About Kennedy Space Center** to access information about all space missions, astronauts, and maps of the area.
- ✎ Check a map and estimate the number of miles you would travel from Disney World to Kennedy Space Center. Which direction would you go? Which highway(s) would you use?
- ✎ Do independent research about other Florida tourist attractions. Make a list of at least five souvenirs and explain where you purchased them.

Florida

Geography

Web site: http://miami.yahoo.com/News/Weather/Miami_Weather/

- Select **Climatology** and compare the current high and low temperatures with record high and low temperatures. Read the precipitation statistics. Is precipitation above or below normal for the month? for the year? What time was sunrise? sunset?
- ✎ Listen to a local weather report on the television. Record the same statistics for your area. Compare the two sets of information.

Web site: http://fcn.state.fl.us/gfc/gfchome.html

- Go to **Wildlife Viewing** and then choose the **Central Region.** What wildlife can you see near Orlando?
- Choose **Viewing By Species.** Read about several species that interest you. Categorize the list into three groups: insects, mammals, birds. Where could you see a manatee?
- ✎ Search the Internet for information about the description, habitat, and feeding of manatee. What is the greatest threat to their safety?

Georgia

General Information

Web site: http://valuecom.com/georgia/info.htm

- Read the information. When did Georgia enter the Union? For whom was the state named? Why does the soil have a reddish tint? Name the five largest cities.
- ✎ What is meant by Confederate States? Make a list of the Confederate States during the Civil War.
- From this extensive site on Georgia history you can read a time line of specific information on a variety of interesting topics.
- ✎ Choose one famous Georgian and write a brief report on his or her accomplishments.
- What are Georgia's leading agricultural and manufacturing products?

Search words: James Oglethorpe, Cumberland Island, Coca-Cola, Jimmy Carter, American Civil War

Georgia

Cities

Web site: http://www.savannah-online.com/travel.asp

- Click on **Travel Guide** and read about several ways to sightsee in Savannah. Click on **Convention and Visitors Bureau** and read about the history of this city.

Web site: http://www.atlhist.org/

- Access **Exhibitions** and view photos of a century of life in Atlanta
- The **Atlanta History Center** is one of the country's largest museums. Take a virtual tour of the museum. Make a list of five items seen at each of these exhibits: Civil War, civil rights, folk arts.
- ✎ What is *Gone with the Wind* and what is its connection to Atlanta?

Web site: http://www.augusta.net/metroaugusta/

- Click on **City Guide** for information on the history of Augusta, then climb aboard the bus for a **virtual tour.**
- ✎ Go to **Maps.** Print a map of the area. Select a major city and trace the most direct driving route to Augusta.

Cut along dashed line.

Georgia

Tourist Attractions

Web site: http://atlanta.yahoo.com/Maps_and_Views/Points_of_Interest/

✎ From this site you can access eleven attractions in Atlanta. Choose one and make a travel poster or magazine cover about it.

- Search the Internet for information about the Atlanta Braves and the Masters Golf Tournament. List the Braves team members last season. Who won the most recent Masters tournament?

Web site: http://www.gacoast.com/navigator/destinations.html

- Welcome to travel opportunities on Georgia's coast. Select **The Golden Isles of Georgia.** Read about **St. Simons Island.** What can you do on a visit to the island?

✎ Choose one of the other coastal islands to compare with St. Simons. Which island is the most remote and natural? Would you enjoy living in this type of environment?

Georgia

Geography

Web site: http://valuecom.com/georgia/info.htm

- Click on **See Map.** What five states border Georgia? What mountains run through the northern section? What is the northernmost city on the map? the southernmost?
- Read the information about **Geographical Regions.**

✎ Draw a map showing the six regions of Georgia. Include the Appalachian Mountains and at least three rivers.

- What are Georgia's most important agricultural products? Which of them does your family use?

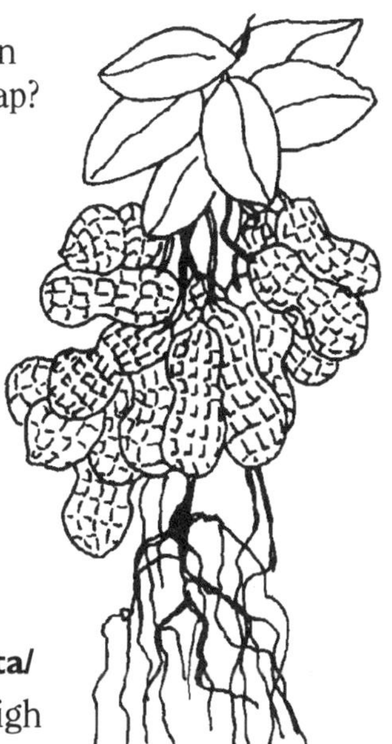

Web site: http://savcvb.com/mapindex.html

- List six public squares on the map of Savannah with their (letter and number) location. Name three streets that run north-south and three that run east-west.

Web site: http://www.city.net/countries/united_states/georgia/atlanta/

✎ Use the information from the **Fact Sheet** to make a graph of Atlanta's high and low temperatures last year.

Hawaii

General Information

Web site: http://www.gohawaii.com/hokeo/school/report.html

- What is the state **tree? bird? marine mammal? fish?**
- ✎ Make a chart showing the **colors and flowers** of each island.
- What is different about the **Hawaiian alphabet?**
- Discuss why the **state flag** is similar to England's flag. Why does it have eight stripes?

Web site: http://hawaii.ivv.nasa.gov/space/hawaii/virtual.field.trips.html

- Select one of the **Virtual Tours,** or go to **Oahu** and take the **Walking Tour of Honolulu.**
- ✎ Use the information to create a travel itinerary for Oahu.

Web site: http://www.global-town.com/history.html

- Click on **Overview** and learn how the islands were formed. Click on the links to read about individual islands.
- ✎ Explain how and why the islands are so dependent on imported goods. What foods are exported? What industry contributes most to the economy?

Search words:
Waikiki Beach, Pearl Harbor, archipelago, Diamond Head, Mauna Loa, Queen Liliuokalani, Mauna Kea

Cut along dashed line.

Hawaii

Cities and Islands

Web site: http://www.co.honolulu.hi.us/

- Who is the **mayor** of Honolulu? What did the mayor do for a living before going into politics?
- What are his six **Goals for a Livable City?** If you were mayor, what would be your goals for a livable city?

Web site: http://www.kauai-hawaii.com/

- This page has many links to the sites and attractions, recreation, parks, and beaches of Kauai. The clickable map will enlarge to give you close-ups and information about individual areas.

Web site: http://www.infomaui.com/

- Use this search engine for **Things to Do** and **Places to See** on Maui. List three locations where a visitor can go snorkeling.
- Go **Shopping** and make a list of local music and food specialties that are available in Maui.

Cut along dashed line.

Hawaii

Tourist Attractions

Web site: http://www.polynesia.com/pcc/Islands/

● The **Polynesian Cultural Center** is one of Hawaii's most popular visitor attractions. It celebrates life and traditions on the seven Polynesian islands. Look at the **Photo Album.** What can tourists do at the center?

● You may e-mail questions about the Pacific Islands from this site.

✎ What kinds of souvenirs could you expect to purchase at the Polynesian Cultural Center?

✎ Name the seven Polynesian islands. Locate them on a map.

Web site: http://www.hshawaii.com/

● This is the home page of the **Hawaiian State Vacation Planner.** You may select one of four islands and search for hotels and restaurants and then book activities.

✎ Suppose you wanted to get married in Hawaii. Plan a virtual wedding. Why do you think so many people go to Hawaii to get married?

Hawaii

Geography

Web site: http://www.bestofhawaii.com/map_oahu.htm

● This site has links to detailed maps of each island. Select **Oahu.** Read the information. Locate Honolulu and Waikiki Beach.

✎ Draw a map of one island. Read information about the geography of that island and create a dimensional model.

Web site: http://www.bestofhawaii.com/weather.htm

● How would you describe the **weather** in Hawaii? What is the temperature range in the islands? What items of clothing will a visitor need?

Web site: http://pubs.usgs.gov/publications/text/Mauna_Loa.html

● View the largest volcanic mountain in the world on the island of Hawaii. When did **Mauna Loa** last erupt?

✎ Do research to learn the way a shield volcano is formed. Explain the process to the class. Locate three other shield volcanoes on a map of the world.

Idaho

General Information

Web site: http://www.state.id.us/other.html

- Search this site for information about Idaho. In what year did it become a state?
- ✎ Study the facts and symbols and make a poster showing the **state seal, flag, bird, tree,** and **flower.**
- Why do you think Idaho has a **state horse** and a state **folk dance?**
- ✎ Use the address or 800 telephone number to request additional information from Idaho's travel council.

Web site: http://www.idahopotatoes.com/

- Visit the home page of the **Idaho Potato Commission.** Click on the **Idaho Potato Difference** and explain why Idaho is perfect for growing potatoes.
- ✎ Who was Luther Burbank and how did he affect Idaho's potato industry?
- ✎ Use **Miscellaneous Potato Facts** to create a bar graph showing how many pounds of each kind of potato the average American eats in a year.

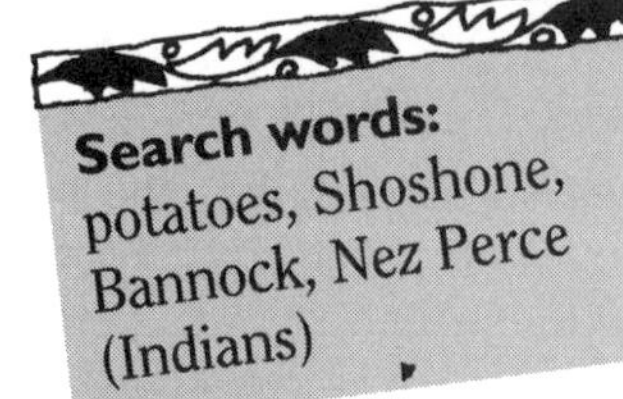

Search words: potatoes, Shoshone, Bannock, Nez Perce (Indians)

Idaho

Cities

Web site: http://www.boise.org/

- Choose **Welcome to Boise** and read about the city's **history.** When and by whom was it founded? What was the Oregon Trail?
- ✎ Search the Internet or draw a map of the Oregon Trail through Idaho.
- ✎ Seven major companies have headquarters in Boise. Search the Internet to find out what each company produces.

Web site: http://www.coeurdalene.com/

- Select **Historical Info,** then click on **Landmarks** for a photo tour of Coeur d'Alene's history. Select **Indians** for a look at the original inhabitants.
- Read what's happening in Coeur d'Alene on this month's **Events Calendar.** List the location for one event you would like to attend.

Web site: http://www.keokee.com/tourguide/guidehm.htm

- From this site you can take a **virtual tour** of Sandpoint, Idaho.
- Use the **Community Profile** to determine Sandpoint's location. How far is it from Canada? What are the Selkirks?

Cut along dashed line.

Idaho

Tourist Attractions

Web site: http://www.visitid.org/

- Read about the **Outdoor Adventures** available for people visiting Idaho. Generally, how would you describe an Idaho vacation? What can you do in the summer? in the winter?
- ✎ Select and order one of the **Free Travel Guide**(s).

Web site: http://www.areaparks.com/index.html

- Locate the **Teton Valley** on a map of Idaho. Label Yellowstone National Park. What three states come together at Yellowstone? Use the area maps to name three cities in the Teton Valley.
- List three reasons why the Teton Valley would or would not be a good place to live.

Web site: http://www.sunvalleyid.com/mall.htm

- From the Sun Valley Online Directory you can access photos of the resort as well as weather reports and information on ski conditions. Use **How to Get There** to locate Sun Valley on a map of the United States.
- ✎ What state borders would you cross traveling the most direct route from your home to Sun Valley?

Idaho

Geography

Web site: http://cirrus.sprl.umich.edu/wxnet/states/idaho.html

- Read the **weather reports** for five cities. Create a chart showing the current temperatures and conditions for those locations. Do any of the cities have weather warnings? If so, include that information on your chart.

Web site: http://www.intellicast.com/weather/pih/radar/

- Check the weather radar for Idaho. Where (and how much) is it raining or snowing? You may need to use another map to understand the abbreviated city names.
- ✎ Locate a weather radar map for your state on the Internet and determine the conditions in three major cities.

Web site: http://www.visitid.org/regions/

- Read about the seven geographical regions of Idaho.
- ✎ Make a map showing all the regions. Draw at least two appropriate icons in each region.
- Choose one region to study in depth. View the map, and write a summary of the region.
- ✎ Draw a regional map to accompany your summary. Label the major cities, rivers, and mountains.

Cut along dashed line.

Illinois

General Information

Web site: http://www.state.il.us/kids/learn/

- The **Just For Kids** site has a wealth of information about the state in a format that is easy to access. You can find the governor's name and take a **photo tour of the capitol.**
- Visit the **Illinois State Symbols Exhibit** and play the **Game.**
- ✎ Create a display of all fourteen Illinois state symbols. Include the lyrics to the state song.
- Go to **History and Geography.** What three U.S. presidents were born in Illinois? Which of the three is most famous? Explain his most important accomplishments.
- ✎ Name ten famous people from Illinois. Tell why each is famous in one sentence.

Web site: http://www.museum.state.il.us/

- Click on **Exhibits,** then **At Home in the Heartland.** This is an exhibit of Illinois history divided into six time periods. You may choose a time frame and review materials from the era.
- ✎ View the **household objects** in each of the sections. Make a time line that shows the items and describe their uses.

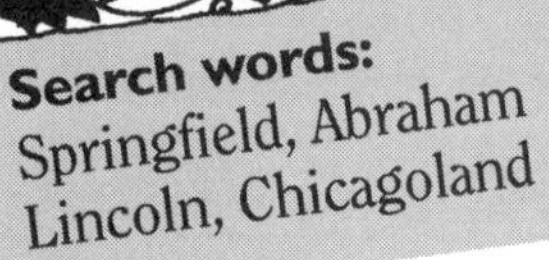

Search words: Springfield, Abraham Lincoln, Chicagoland

Illinois

Cities

Web site: http://www.ci.chi.il.us/

- Who is the mayor of Chicago? Select the **Electronic Tour Guide** and make a list of **Things to Do and See** in Chicago.
- Use the **Interactive Map** to visit Michigan Avenue, called the Magnificent Mile. List three reasons why it has been given that nickname.

Web site: http://www.villageprofile.com/peoria/

- Learn about the **History** of the city of Peoria. Name two early settlers. Give three examples of early industries in the area. What was produced by the Benjamin Holt Company?
- Go to **Art and Entertainment.** What performances and museums are available for art and music lovers in Peoria?
- ✎ Click on **Location.** What is Peoria's geographic location in Illinois? Draw a map and label Peoria as well as one city to the north, south, east, and west.

Cut along dashed line.

Illinois

Tourist Attractions

Web site: http://www.enjoyillinois.com/

- Plan a getaway using the **Illinois Trip Planner.** Follow the simple directions to choose a category and destination that appeal to you. Read the choices and answer a few questions. Print your results.

Web site: http://www.nps.gov/liho/

- What will a visitor see while touring Lincoln's home? What is the entry fee? What are the hours of operation?
- ✎ Search the Internet for a biography of Abraham Lincoln. Whom did he marry? What were his children's names? What years was he president of the United States? How did he die?

Web site: http://www.villageprofile.com/vp1.html

- Use the **Community Directory** or graphics to select and access many Illinois communities.
- ✎ List the current mayors of ten cities. Read their welcome messages. Make a chart showing the city, mayor's name, and two attractions in each community.

Illinois

Geography

Web site: http://www.commerce.state.il.us/dcca/files/kids/heritage.htm

- Name the five states that border Illinois.
- What is the largest inland lake?
- ✎ Name any eight major Illinois rivers and label them on a map of the state.

Web site: http://www.villageprofile.com/locator.html

- Use the active links on the **Illinois Locator Map** to visit counties and cities of your choice.

Web site: http://www.commerce.state.il.us/dcca/menus/coalcool.htm

- Illinois ranks fifth among coal-producing states in the United States. What are some problems associated with mining and burning coal? How does Illinois intend to solve those problems? Why does coal mining continue to be important to the state?
- ✎ Check the air pollution regulations in your area that might restrict the burning of coal.
- ✎ What is meant by land reclamation? Why is it important to communities?

Cut along dashed line.

Indiana

General Information

Web sites: http://www.state.in.us/sic/index.html
http://www.state.in.us/sic/HTML/general_facts.html

- Click on **Little Hoosiers–Kid's Page** and find out why the people of Indiana are called Hoosiers. What is the state flower? tree? bird? stone? river?
- What is pictured on the state seal?
- What is the state motto?

Web site: http://www.nps.gov/gero/

- Learn about the history of Indiana during a visit to **George Rogers Clark National Park** at Vincennes. Who was George Rogers Clark?
- ✎ Search the National Park Service site for two other Indiana parks.

Web site: http://city.net/countries/united_states/indiana/

- ✎ Use the temperatures in **Today's Weather** to make a chart showing high and low temperatures in Fahrenheit and Celsius for Evansville, Fort Wayne, Indianapolis, and South Bend.

Search words:
Indy 500, Brown County, Gary, Indiana Dunes National Lakeshore, Purdue University

Indiana

Cities

Web site: http://www.indplsconnect.com

- From this site you can access Indianapolis **sites, attractions,** and **things to do.** If you could visit only one **Museum,** which would you choose? Explain.
- Go to **Attractions,** then visit the **Virtual Zoo** and read the **Animal Questions.**
- ✎ Write a question you would like answered about your favorite zoo animal. Send an e-mail to zoo personnel.

Web site: http://www.lafayette-online.com/

- Take a **virtual tour** of the city and business district of Lafayette. Read the **History** of Lafayette. What geographic features helped the city prosper? Name three major manufacturers.
- ✎ Write about the historical significance of the Battle of Tippecanoe.

Web site: http://www.evansville.net/eville/profile/

- Click on **Area Statistics.** What counties are included in the Evansville area? How does the size (in square miles) of the city compare to the size of the multicounty area?
- ✎ Choose a city from the list of **Distances to Major Cities.** Trace a route to Evansville on a map of the United States. What states would you pass through to get there?
- ✎ Use the **Climate** information to make a line graph of monthly average temperatures.

Cut along dashed line.

Indiana

Tourist Attractions

Web site: http://www.indyracingleague.com/

- Select **The Indianapolis Motor Speedway.** Go to **Stats** and read the list of **Indy 500 Winners.** What is displayed at the Hall of Fame **Museum?**
- Click on **The Indy 500** and view the **Photo Gallery.**
- ✎ Choose a recent Indy 500 winner. Locate his statistics in these links.

Web site: http://www.childrensmuseum.org/

- Explore the world's largest children's museum. Check out the pictures in **What's Cool.** Write a description of three exhibits. **Plan a Trip** to the museum. What are the hours and admission fees?
- ✎ Print a **floor plan** of the museum. Locate and label five **online exhibits.**

Web site: http://collegefootball.org/tour/

- The **College Football Hall of Fame** is located in South Bend, Indiana.
- ✎ Make a map of the hall showing five exhibits. Summarize what you can see and do at each one.

Indiana

Geography

Web site: http://www.state.in.us/dnr/

- Go to **Historic Preservation and Archaeology.** What four **criteria** must an old building or archaeological site fulfill to be placed on the National Register? Why is it important to reuse and preserve historic sites?

Web site: http://www.state.in.us/sic/HTML/mileage.html

- ✎ Select three Indiana cities from the top of the chart. Locate the city nearest to your home from the list on the left side of the chart. Estimate how long it would take to drive to each of the Indiana cities and how much gasoline you would use if you averaged fifty-five miles per hour and thirty miles per gallon.

Web site: http://www.cica.indiana.edu/news/servers/tourist/index.html

- Locate Indianapolis on the map. Which direction would you drive from Indianapolis to reach each of these cities: Hanover, Lafayette, Kokomo, South Bend, and Evansville?

Iowa

General Information

Web site: http://www.silosandsmokestacks.org/

- This site has interesting information about the cultural heritage of Iowans. From what countries did they come? How did these people help to establish Iowa as an agricultural state?

Web site: http://www.state.ia.us/tourism/index.html

- Go to **Fun Facts.** What two rivers border Iowa? Explain Iowa's importance as an agricultural state.
- ✎ Make a graph showing the population of Iowa's top ten cities.

Web site: http://iowa-counties.com/agriculture/

- Go to **Beef Cattle,** then **Beef Facts: Byproducts** for interesting information. Explain how the industry uses 99 percent of each animal.
- Select **Grains** and go to the **Maize Page.** Use the **Maps** to estimate Iowa's corn production per acre. Learn **How a Corn Plant Develops** from the agricultural site at the University of Iowa.

Search words: Buffalo Bill Cody, Black Hawk Indians, Mississippi River, Missouri River, maize

Iowa

Cities

Web site: http://www.desmoinesia.com/Info.html

- Read the **History** of Des Moines. What was its original name? For what purpose was the original military post established?
- What kinds of businesses are headquartered in Des Moines?
- Go to **Attractions.** What "simple pleasures" are available for family adventures?

Web site: http://www.ames.ia.us/

- Go to the **Convention and Visitors Bureau.** What highways lead to Ames? What university is located there?
- Select **Things to Do.** This listing is for Ames and the surrounding communities. What first lady was born in Boone, Iowa? What attractions can be found in Story City?
- Print the **Map of Ames.** What roads border the University of Iowa? Which direction would you travel to visit Story City? to visit Des Moines?
- ✎ Describe in words the route from the university to the airport.

Cut along dashed line.

Cut along dashed line.

Iowa

Tourist Attractions

Web site: http://www.state.ia.us/tourism/index.html

- Go to the **top ten Iowa vacation areas.**

✎ Use this interactive map to make your own map of Iowa's vacation spots.

Web site: http://www.quadcities.com/cvb/

- Click on **Introduction.** Describe the communities that make up the Quad Cities. Why is the area unique? Why is it a good tourist destination? What can you do on the **Mississippi River?**
- Select **History and Museums.** What would you see on a two-day tour of the area's museums?

Web site: http://www.ioweb.com/lhf/index.html

- Visit the **Living History Farms,** an open-air agricultural museum. Visit re-creations of an American Indian village, working farms, and the town of Walnut Hill.

✎ Request a **Visitor's Packet** by e-mail.

Web site: http://www.5seasons.com/

- Check out the **Five Seasons Center** in Cedar Rapids, Iowa. List some coming attractions.

Iowa

Geography

Web site: http://www.icvba.org/

- Study the highway map of Iowa. What two highways cross at Des Moines? How many miles is it from Des Moines to Council Bluffs? Name four cities on the eastern border of the state.

Web site: http://www.profiles.iastate.edu/places/

- Use the interactive **county map** to select and compare two cities within the same county, or one city from each of two counties.

✎ Draw a detailed map of one county and label cities and highways. Indicate the counties immediately to the north, south, east, and west.

Web site: http://www.nativenations.com/iowa.html

- List the **Original Nations of Iowa (before 1700).** Go to **Ioway.** What is the meaning of the word *Iowa?*

✎ Read about the Ioway nation. Make a chart describing its history for three periods: prehistory–1685, 1685–1838, and 1838–present. Include how and where they live today.

- Search the archaeological area of the **Blood Run National Historic Landmark.**
- Describe the purpose and appearance of the earth mounds built by the American Indians in the Late Woodland period.

Kansas

General Information

Web site: http://lawlib.wuacc.edu/washlaw/kansas/kansas.html

- Select **Heritage and History,** then **Kansas Forts Network.** Read about the history of Kansas forts and a typical soldier's life.
- ✎ Draw a map of Kansas and label the eight forts.

Web site: http://www.go-explore.com/GetCategories?stateList=Kansas

- Read about the **History** of Kansas. Who were the first inhabitants? Natives of what European countries settled the area? What group of immigrants introduced wheat production?

Web site: http://www.ksu.edu/wheatpage/funfacts.html

- Go to **Kansas Average Wheat Yields Per Acre.** Which year had the highest yield per acre? the lowest?
- Read **Facts about Kansas Wheat.** How might the United States be affected if Kansas had a drought or flood? What jobs (other than farming) are created because of Kansas wheat production?
- ✎ Explain the nutritional value of a kernel of wheat.

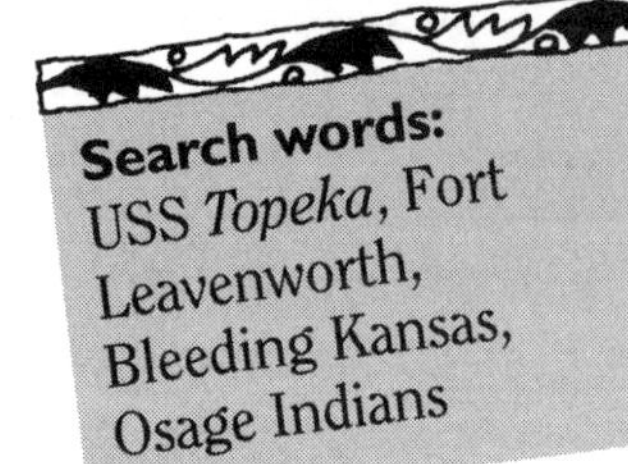

Search words:
USS *Topeka*, Fort Leavenworth, Bleeding Kansas, Osage Indians

Kansas

Cities

Web site: http://lawlib.wuacc.edu/washlaw/kansas/cities.html

- From this site you can access information about many Kansas cities.
- ✎ Choose two small cities and compare housing, shopping, dining, and business opportunities.
- ✎ Search for information about several small Kansas towns on the Internet. Choose the best one in which to live and work. Explain your choice.

Web site: http://www.topeka.org/

- From this site you can access information about Topeka, the state capital. Use the pull-down menus for quick connections to **government** and **attractions.** View **photos** of the city.
- What are the current weather conditions in Topeka?

Web sites: http://leavenworth-net.com/
http://www-leav.army.mil/cac/history.htm

- Fort Leavenworth was Kansas's first city. Read **History and Tour.** Why was the fort established? What is on display at the **Army Museum?**

Cut along dashed line.

Kansas

Tourist Attractions

Web site: http://www.kcmuseum.com/index01.html

- At this site you can enter the **Kansas City Museum** or **Science City at Union Station.** Choose one of the following activities to complete:
- ✎ Go to **History** and then **Native Lands.** Briefly explain the history and lifeways of the Osage American Indians. Copy the map showing their territory.
- ✎ Complete a time line with ten events from the **History of Union Station.**
- ✎ Make an original drawing of three exhibits seen during a **walk-through** of Science City.

Web sites: http://www.dodgecity.net/info/history.html
http://banzai.neosoft.com/citylink/dodge/default.html

- Read about the history of Dodge City. What makes it a popular tourist destination? What major highways lead to the city? Who is the mayor?
- ✎ Make a list of Dodge City's historical sites.

Kansas

Geography

Web site: http://gisdasc.kgs.ukans.edu/kanview/landcov/landcover.html

- Click on three counties and read listings of their **land cover.**
- ✎ Make a chart showing counties that have cropland, grassland, water, and woodland. What covers most of the state?

Web site: http://gisdasc.kgs.ukans.edu/dasc/city/ks_city.html

- Use the **interactive city map** to read demographics for the city of your choice. Choose five cities.
- ✎ Estimate the male and female population numbers by rounding to the nearest thousand and show all the statistics on a bar graph.

Web site: http://www.npca.org/npt/zbar.html

- Learn about the **Tallgrass Prairie National Preserve.** What buildings stand on the 115-year-old ranch? Read the **Tour on the Prairie** and describe how the prairie land was formed in prehistory. For what is the land best suited?
- Read about the **current status** of the prairie. What legislation protects the area?
- ✎ Explain the prairie ecosystem, including the effects of wildfires, grazing, and weather.

Kentucky

General Information

Web site: http://www.state.ky.us/tour/kentucky_facts.htm

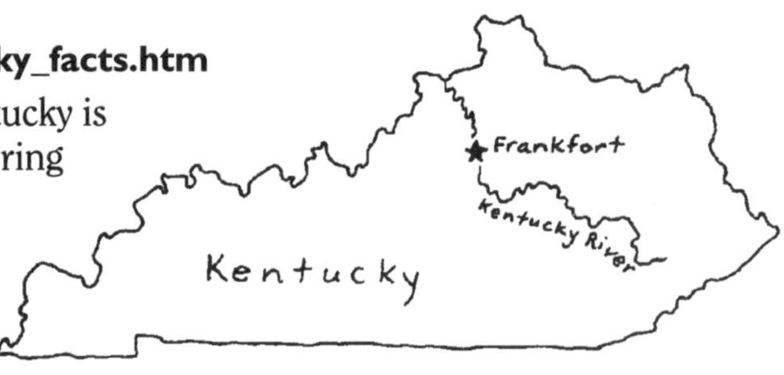

- Read the information to determine where Kentucky is located in the United States. Name seven bordering states. What river occupies the northern and western borders?
- How important is agriculture to Kentucky's economy? What are the chief products? What minerals are produced?

Web site: http://www.state.ky.us/tour/kentucky_facts.htm

- Read the lists of **Famous Kentuckians.**
- ✎ Search the Internet for additional information about one person who interests you. Write a brief biography of the person explaining his or her accomplishments.

Web site: http://www.state.ky.us/agencies/gov/symbols.htm

- Why is Kentucky called a commonwealth? What other states are commonwealths?
- ✎ Make an illustrated chart of Kentucky's symbols.

Search words:
Bluegrass, Kentucky Horse Park, Lexington

Cut along dashed line.

Kentucky

Cities

Web site: http://www.frankfortky.org/

- Read **About Frankfort.** Where is it located within the state? Why was it chosen Kentucky's capital? If you were **Moving to Frankfort,** what cultural and recreational resources are available?

Web site: http://www.louisville.com/

- Select **Community** for a long list of city guides and links. If you were moving to Louisville, which neighborhood would you choose? Explain.
- Click on **Museums.** Where is the Colonel Harland Sanders Museum located? What will you see there? Where can you get information about the Louisville Slugger baseball bat?
- ✎ When is the Kentucky Derby? What are some of the traditions of this race? Who was the most recent winner?

Web site: http://www.tourky.com/williamsburg/information.htm

- Read the **General Information** about Williamsburg. What is unique about its location? Search all the sites carefully. What efforts is the city making to attract tourists? Why would the tourist business be important to this city?
- ✎ Who was Daniel Boone? Why would a national forest be named for him?

Cut along dashed line.

Kentucky

Tourist Attractions

Web site: http://www.state.ky.us/agencies/parks/parki65.htm

- This site has extensive information about **Kentucky's State Parks.** In which region could you go caving? see waterfalls? climb mountains?
- ✎ Choose a region of the state and study the parks and historic monuments. Give details about the attractions and accommodations available to visitors. Draw a sketch map of the route from your home to the region of Kentucky where the park is located.

Web site: http://www.imh.org/khp/

- Go to **General Information** about the Kentucky Horse Park. How many breeds of horses live at the park? What can a visitor see at the **International Museum of the Horse?** People from which European country brought the horse to America?
- ✎ Read about Man o' War. Why is he considered to be one of the greatest racehorses ever?

Web site: http://www.nps.gov/maca/macahome.htm

- From this site you can explore Mammoth Cave, the most extensively known cave system on earth. Go to **Mammoth Cave in Depth** for a look at the plants, animals, and geology of the park.
- ✎ Read **Archaeology** and create a time line explaining the lives and activities of people in the area since prehistory.

Kentucky

Geography

Web site: http://www.state.ky.us/tour/maps/distance.htm

- Use the map and chart to determine the distance traveled on a Kentucky vacation from Frankfort to Louisville to Bowling Green. Name three cities you would pass if you traveled across southern Kentucky from Hazard to Murray.
- How far is the state capital, Frankfort, from Memphis, Tennessee? from St. Louis, Missouri? Which is farther from Detroit, Michigan: Louisville or Bowling Green?

Web site: http://www.fb.com/kyfb/roadside.htm

- Kentucky is an agricultural state. This map shows the **Roadside Farm Markets** that are popular throughout Kentucky. Which region of the state has few markets? Why is that?
- ✎ Read the **directory** and make a chart showing the products available during each season.

Louisiana

General Information

Web site: http://www.state.la.us/state/student.htm

- Click on **Profiles,** then **People.** People from what European countries colonized Louisiana?
- What is meant by **Cajun Country?** Locate the area once known as Acadia on a map of the world. In what ways is "cajun" a part of Louisiana's culture?
- ✎ Search the sites for evidence of the European heritage that remains in the culture of Louisiana and New Orleans.

Web site: http://www.crt.state.la.us/crt/sayit.htm

- ✎ Use the word list to make a pictionary of ten special words spoken by people living in Louisiana.

Web sites: http://www.crt.state.la.us/crt/flags.htm
http://www.sec.state.la.us/brief-1.htm

- ✎ Draw a time line showing five Louisiana flags.

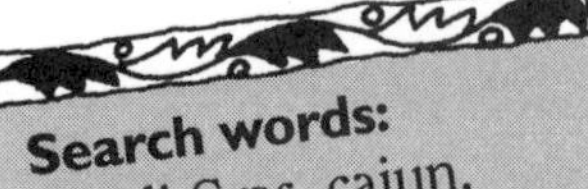

Search words:
Mardi Gras, cajun, creole, Louisiana Purchase, zydeco, plantation, parish

Cut along dashed line.

Louisiana

Cities

Web sites: http://www.brnet.com/index.html
http://www.pipeline.com/~baton.rouge/

- Go to **Government** and search the links for the name and address of the governor.
- Learn current **weather** conditions and read the headlines from the *Advocate,* a local newspaper.
- View a live cam image of downtown.
- ✎ Go to **Recreation and Tourism,** then **Local Attractions.** Write an itinerary for a three-day visit to the city. Print a **map.**

Web site: http://www.experienceneworleans.com/

- Click on **Sights** to learn about **The Cities of the Dead,** a restored **plantation,** and a **Louisiana Swamp.**
- Why are tombs above ground in New Orleans cemeteries?
- What makes up the walls of these cemeteries?
- What kinds of work had to be done to restore the plantation? Discuss the pros and cons of restoring a historic home or building a new one. Which would you do? Explain.

Cut along dashed line.

Louisiana

Tourist Attractions

Web site: http://www.crt.state.la.us/crt/sbcover.htm

- Select **Mardi Gras** and explain the religious background of the celebration. In what cities is the old European tradition still followed?

Web sites: http://www.experienceneworleans.com/
http://www.honeyislandswamp.com/

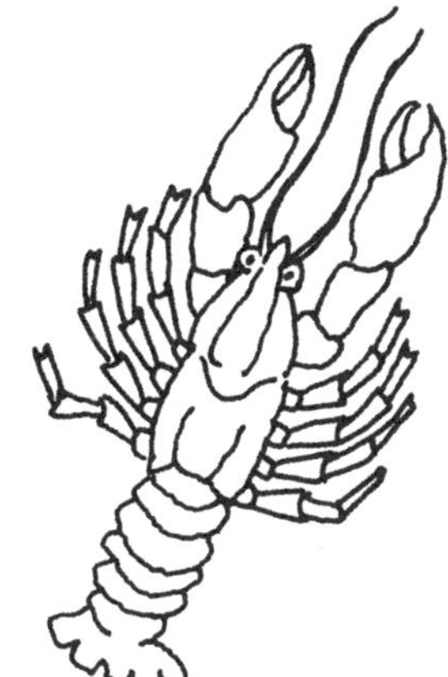

- What is a swamp? a bayou? Why is it important to preserve **Louisiana's Honey Island Swamp?** What did you learn about the plants and animals living there? What would you see and hear on a tour of the swamp?

Web site: http://www.superdome.com/tours.html

- Click on **Tours** for details about the Superdome's size and history. What record was set there in 1981?
- What athletic teams play at the dome? Check the **Photo Gallery.** What other kinds of performances are held in the Superdome?
- ✎ Why is New Orleans a good location for the Superdome? If you were touring Louisiana, would you make an effort to attend an event at the Superdome? Explain.

Louisiana

Geography

Web site: http://www.crt.state.la.us/crt/purch.htm

- Explain the history of the **Louisiana Purchase.** How did it affect the size of the existing United States? What famous leader sold it to the United States? Who was the U.S. president at the time?
- ✎ Draw a map showing the thirteen states that would be formed from the Louisiana Purchase.

Web site: http://www.crt.state.la.us/crt/sbregion.htm

- Read about each of the five **regions** of Louisiana.
- ✎ Make a large map of the state and label the regions. Indicate the native countries of the original settlers and draw an icon to symbolize the culture of each region.

Web sites: http://www.weather.com/weather/us/states/Louisiana.html
http:// www.intellicast.com/weather/btr/
http://www.lanyap.com/no/weather.htm

- Compare the weather conditions in two different areas of Louisiana. Make three general statements about the climate. If you were moving to the state, name three items of clothing you would probably not need to take along.

Maine

General Information

Web site: http://www.state.me.us/mema/weather/genweath.htm

- Select **State Forecast** and use the information to design icons for a weather map of Maine.

Web site: http://www.visitmaine.com/facts.html

- What is the state's nickname? gem? cat? insect? tree? bird?

Web site: http://www.bathmaine.com/

- This museum preserves and displays the maritime history of Maine. Where is the museum located? What is currently on **Exhibition?** What kinds of skills are demonstrated at the museum?
- ✎ What jobs are unique in a coastal location like Maine?
- ✎ What is a schooner? Search the Internet for information and draw a diagram.

Search words:
Gulf Hagas, Mount Desert Island, Isle au Haut, Schoodic Peninsula, Kennebunkport, Portland

Maine

Cities

Web site: http://www.maineguide.com/

- Select **Explore Maine,** then choose **Bangor** from the pull-down menu. Take a **Walking Tour of Downtown.** Explain the history of the Bangor House. What is the derivation of the word *Norumbega?*
- ✎ Print the map of Downtown Bangor and use the addresses to trace the walking tour.

Web site: http://www.maineguide.com/

- Select **Explore Maine,** then choose **Augusta.** Read the historic information about this city. When was Maine admitted to the Union? Make a list of historic buildings and museums.
- Click on **Maine's Government.** Who is Maine's governor?

Web site: http://www.maineguide.com/

- Select **Explore Maine,** then choose **Moosehead Lake Region.** Read about the **Mountain Hikes.** What precautions should be taken before beginning one of these hikes? Why is Gulf Hagas special?
- Read the **Scenic Tours.** How does life change for residents of Moosehead with the change of seasons?

Cut along dashed line.

Maine

Tourist Attractions

Web sites: http://www.gorp.com/gorp/resource/US_National_Park/me_acadi.HTM
http://barharbor.com/acadia.html
http://www.americanparknetwork.com/parkinfo/ac/

- View photos and read information about **Acadia National Park.** What is the highest point in the park? What activities are possible?
- Click on **History.** How did John D. Rockefeller, Jr., contribute to the development of Acadia and Mount Desert Island? What disaster occurred in 1947?
- ✎ Explain the rules about trash disposal that must be observed by campers on the Isle au Haut. Give a reason for those rules. Why aren't pets allowed? Why is biking discouraged?

Web site: http://www.visitmaine.com/lighthouse.html

- Explain why **Lighthouses** were once so important to Maine's coast. For what are they used today?
- ✎ Read the directory and locate ten of the lighthouses on a map.
- Go to **Covered Bridges.** How many of these bridges remain in Maine? Why were they built? How were they destroyed?
- ✎ Locate and label the bridges on a highway map.

Maine

Geography

Web site: http://www.americanparknetwork.com/parkinfo/ac/

- Select **Geology.** How did glaciers help shape the lakes and cliffs of the region? How many individual mountains were formed?
- Where did the glaciers originate?
- Go to **Preservation.** Why is it important to maintain Acadia's "wildness"? What is being done to preserve areas at risk?

Web site: http://www.state.me.us/doc/prkslnds/map.htm

- Read the map key. Name a state park and historic site in the southern and eastern parts of the state. Name three public reserved lands in the north and west.
- ✎ Compare this map with a city map. Name the park or historic site nearest each of these cities: Augusta, Bangor, and Bar Harbor.

Maryland

General Information

Web site: http://www.mdarchives.state.md.us/

- Select **Maryland and Its Government** and go to All Ab**out Maryland.** Read the biography of Maryland's current Governor. What political offices has he or she held? What are his or her goals as governor?
- Read **All About Maryland**. For whom was Maryland named? What is the population of the state? How has its proximity to Washington, D.C., affected Maryland's economy and population growth?
- What is the state sport? boat? dog? flower? tree?
- ✎ Make a drawing of Maryland's **state flag** and explain its symbolism.

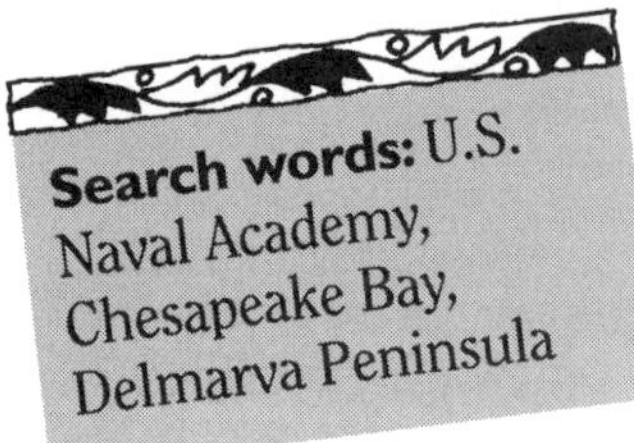

Search words: U.S. Naval Academy, Chesapeake Bay, Delmarva Peninsula

Cut along dashed line.

Maryland

Cities

Web site: http://www.mec.state.md.us/meccity.html

- Use this list to access information about Maryland's cities. Select **Annapolis** and go to **Virtual Tour.** Click on **Index to Key Sites** and read the brief information.
- Select **Baltimore** and go to **Virtual Tour.**
- ✎ Make a chart that summarizes the attractions offered in each neighborhood of Baltimore.
- Select **Germantown.** Go to **Statistics,** then **Age.** Which age group makes up most of Germantown's population?
- Go to the **Chamber of Commerce.** What is the **Master Plan** for the city? By how much are the population and employment base expected to increase by the year 2000?

Web site: http://city.net/img/magellan/mgmapsoftheworld/baltim.gif

- ✎ Print the map of Baltimore. Label the neighborhoods.

Cut along dashed line.

Maryland

Tourist Attractions

Web site: http://www.nps.gov/anti/

- Go to the **Virtual Visitors' Center.** Read **Battlefield Information.**
- ✎ Explain in detail the historic significance of the battle at Antietam.

Web site: http://www.beach-net.com/

- What can you do for free on the beach? Which of those things would you want to do?
- What kinds of museums are located in Maryland's beach area?
- ✎ Click on **Directions.** Choose a state and city a day away. Mark the beach and the starting point on a map. Trace the route.

Maryland

Geography

Web site: http://www.mdisfun.org/kids/

- Go to **Geography.** What states border Maryland? What is the highest point? Where is the lowest point? How many miles of coastline border the Chesapeake Bay?

Web site: http://www.capitalonline.com/tour/annapmap.html

- Who laid out the streets in the city of Annapolis? Name two streets that border the U.S. Naval Academy. In what part of the city is the academy located? Where are the city dock and Spa Creek?

Web site: http://www.chesbaynet.com/

- Read the **Introduction to Chesapeake Bay.** What cities are located on the bay? What is the meaning of the word *Chesapeake?* Organize the **Points of Interest** according to location: upper, middle, or lower bay.
- ✎ How long would your family need to explore Chesapeake Bay? Which would you probably choose to do first? Explain.
- ✎ What is an estuary?

Cut along dashed line.

Massachusetts

General Information

Web site: http://www.magnet.state.ma.us/sec/cis/cismaf/mafidx.htm

- Select **Massachusetts Profile.** When did Massachusetts become a state? What is the meaning of the state motto considering early American history?
- ✎ Name the original thirteen states. Label them on a U.S. map.
- Select **State Symbols.** Where did the name *Massachusetts* originate? What are the state's nicknames?
- ✎ Explain the symbolism on the state seal and the state flag.

Web site: http://www.magnet.state.ma.us/sec/cis/cismaf/mf4.htm

- Read the list of **Famous Firsts** in Massachusetts.
- ✎ Choose six items that interest you to represent on an illustrated time line.
- Which four U.S. **presidents** were born in Massachusetts?
- ✎ Do research to find the years that each one held office. List one accomplishment for each.

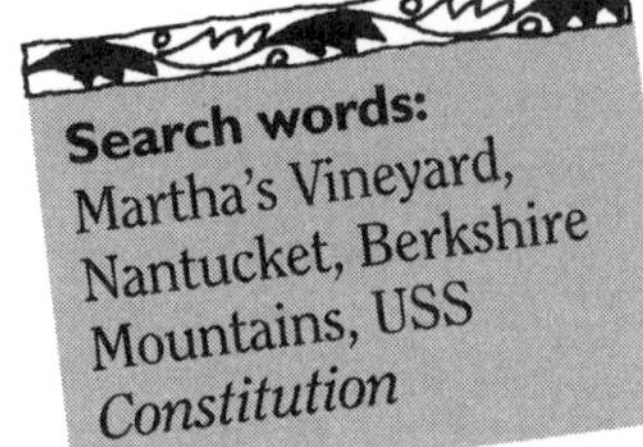

Search words: Martha's Vineyard, Nantucket, Berkshire Mountains, USS *Constitution*

Massachusetts

Cities

Web sites: http://www.bweb.com/bostonweb/index.htm
http://boston.yahoo.com/

- Click on **Walking Tour** and trace Boston's Freedom Trail. What famous Americans are buried at the Granary Burying Ground?
- View the **Photo Tour** for pictures of these historic landmarks.
- ✎ Read more information about Paul Revere and the importance of the Old North Church to American history. Read "Paul Revere's Ride" by Henry Wadsworth Longfellow.
- ✎ Explain the historic significance of the Boston Tea Party.

Web site: http://www.ktiworld.com/weather/KTI-Amherst-weathercam.html

- From this site you can see a live view of a street in Amherst, Massachusetts. Click on the links and take a walking tour of historic homes.
- Print the **Street map** and locate these sites.
- ✎ When did Emily Dickinson live? Research the titles of three of her poems.

Cut along dashed line.

Massachusetts

Tourist Attractions

Web site: http://www.ici.net/cust_pages/jack/mamie.html

- Visit the five ships on display at **Battleship Cove.** Name them and identify the war in which each one served. What is *Big Mamie?* What was significant about its service during World War II?
- At what nearby museum can you find artifacts and information about the *Titanic?*
- ✎ Explain the story of the *Titanic.*

Web sites: http://media3.com/pilgrimhall/
http://media3.com/plymouth/history/

- In what part of Massachusetts is Plymouth located? What remnants of the early Pilgrims can be seen there?
- ✎ Read the information and explain in detail why the Pilgrims came to America.
- This is the site of **Pilgrim Museum,** the oldest museum in continuous operation in America. Look at the photos and read the information. Think of five other Pilgrim artifacts that might be displayed in the museum.

Massachusetts

Geography

Web site: http://www.magnet.state.ma.us/sec/cis/cismaf/mf1c.htm

- Select **Statistics.** What is the area (in square miles) of the state? What is its biggest county?
- How much of the state is populated by immigrants? From which countries have they come?
- Go to **Boundaries.** What is the geographic location of Massachusetts? What states occupy its borders?
- What is the general topography of the state? Where is the best *soil* for farming? Which berries grow well in the Cape Cod area?
- Go to **Islands.** Which islands are popular summer resorts?

Web site: http://www.virtualcapecod.com/facts-g.html

- What is the topography of Cape Cod?
- Select **Beaches on Cape Cod.** Take the **Virtual Tour.** Select a city, view the map, and go to **Town Facts.** Describe the city in terms of activities, points of interest, and beaches.

Michigan

General Information

Web site: http://www.migov.state.mi.us/

- Click on **State Capitol.** What is unique about Michigan's capitol building? When was it renovated?
- Who is Michigan's governor? What other state political offices has he held? Name the first lady.

Web site: http://www.sos.state.mi.us/history/michinfo/michinfo.html

- Summarize information about these dates in Michigan's history on an illustrated time line.
- ✎ Read **March 6, 1896,** and explain in detail the people involved in car manufacturing and development in Detroit. Search the Internet for photos of historic automobiles. Print and label them.
- Go to **State Symbols.** Why was the apple blossom selected as Michigan's state flower? Why is the painted turtle Michigan's state reptile?
- Go to **Famous Michiganians.** Choose one and search the Internet for additional biographical information.

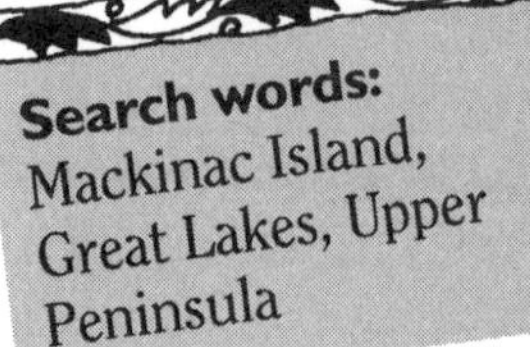

Search words:
Mackinac Island, Great Lakes, Upper Peninsula

Michigan

Cities

Web sites: http://www.ci.ann-arbor.mi.us/
http://www.sils.umich.edu/AnnArbor/

- Enter either site for the city of **Ann Arbor.** Make a list of items that are of interest to citizens of the city. Read the **History of Ann Arbor.** Who were its founders? How did it get its name?
- Tour the sites by viewing a University of Michigan student's postcards home. Read the information and explain the university's importance to the community.
- ✎ Design two postcards for landmarks in your city. Include a message about each site.

Web site: http://detroit.freenet.org/gdfn/detroit/detroit.html

- Read the **General Facts about Detroit.** Where is the city located? How many counties does Greater Detroit include? What is their combined population?
- What are Detroit's major industries other than automobile manufacturing?

Web site: http://www.frankenmuth.org/

- Check out Michigan's number one tourist destination, called Little Bavaria. Before you look at the site, list three things you could expect to find in Frankenmuth.
- Go to **History.** What does the name *Frankenmuth* mean in German? Who were the original settlers of this town?

Cut along dashed line.

Michigan

Tourist Attractions

Web site: http://www.hfmgv.org/about.html

- What categories of collections are housed at the **Henry Ford Museum?** What historic structures have been relocated to **Greenfield Village?**
- Click on **Museum and Village Highlights.** Look at several examples.
- ✎ Choose one category from each site and search the Internet for details about the history associated with the items exhibited.

Web site: http://www.citysidewalks.com/

- Click on the icon to enter this site. Go to **Museums** and then visit the **Motor Sports Hall of Fame.** Why do you think this museum is located in Detroit?
- ✎ Under what categories are Hall of Fame members listed? Choose one member. Search the Internet to find the statistics that qualified him or her for this honor.

Web site: http://www.portup.com/traveler/home.html

- What tourist attractions are available on Upper Michigan's **Keweenaw** Peninsula? Search the site for an activity to do in each season of the year. Explain the history of the **Douglass House.** Why was it built? How has it changed since 1860?

Michigan

Geography

Web site: http://www.nps.gov/isro/

- Locate **Isle Royale National Park** on a map of Michigan. In which Great Lake is it located? What outdoor activities are done in the park? What is the climate of the park? What kinds of clothing are necessary?

Web site: http://www.nps.gov/kewe/

- What metal was mined at **Keweenaw?** How did that metal affect the industrial development of the United States? What special services are offered to visitors to this historical park?
- ✎ Search the Internet for the names of two national lakeshores. What are dunes?

Web site: http://www.portup.com/traveler/home.html

- Select **Moose on Isle Royale.** Read what is occurring in the **Wolf and Moose Study** at Isle Royale National Park. Explain the importance of the predator-prey relationship in maintaining both species. What could be done to resolve the situation that exists?
- Click on **About Bears.** In which season are bears the most active? Explain. What should campers or hikers do if they see a bear? Why is it unwise to feed bears in a park setting?

Minnesota

General Information

Web site: http://www.state.mn.us/aam/

- Select **Sky Blue Water** from the **Table of Contents.** What American Indian group gave Minnesota its name? What created all the lakes and rivers? What are Minnesota's five largest cities?
- What are **Minnesota's Major Industries?**
- What is Minnesota's bird? tree? muffin? drink?
- Why does Minnesota have a state grain?

Web site: http://deckernet.com/minn/

- At this site you can read the headlines or weather report for the day. What are the current weather conditions in Minnesota? If possible, **Listen to the Loon.**

Web site: http://www.tccom.com/amateur/news.html

- Read the information about **Amateur Athletes in Minnesota** who hope to go to the next Olympics. Select your favorite athlete or sport and click on **continued** to read the **interviews.**
- ✎ Click on **Mail** to send e-mail to an athlete from this site.

Search words:
Twin Cities, Hubert Humphrey, loon, Paul Bunyan, Chippewa Indians, Laurel Indians

Minnesota

Cities

Web site: http://www.city.net/countries/united_states/minnesota/saint_paul/

- Use the pull-down menus in **City of St. Paul** to answer questions about government and entertainment.
- ✎ Select **Fun Activities** for information about the most current attractions. List three events that you would like to attend.

Web site: http://www.minneapolis.org/

- Click on **Discover Minneapolis** for **Fast Facts** about the business community. What Fortune 500 companies have headquarters there? What products have been invented by Minnesotans?
- Continue to **Fun Facts and Trivia.**
- ✎ What sporting events have been held at the Hubert Humphrey Metrodome? Design a poster or T-shirt logo advertising one of the events.
- ✎ Fill in the form to request a **Twin Cities Official Visitor's Guide.**

Web site: http://www.intlfalls.org/index.htm

- Where is **International Falls** located? What national park is there?
- Select **Attractions.** What football legend is honored with a museum? Members of what American Indian tribe are buried in Grand Mound? Who was Chief Woodenfrog?

Cut along dashed line.

Minnesota

Tourist Attractions

Web site: http://www.mnhs.org/histctr/histctr.html

- Review the exhibits at the **History Center.** Write six facts you learned about life in Minnesota from these **Exhibits.**
- ✎ Work with a group to write a list for your state (or another state you have studied) similar to **Minnesota A–Z.**

Web site: http://www.bluffcountry.com/

- In what region of Minnesota is **Bluff Country** located? Name five nearby communities. What kinds of lodging and food are typical of the area? View the **Photo Tour** to sample the natural beauty.

Web site: http://www.mnhs.org/prepast/mnshpo/larger/larger4.html

- This clever site has photos of **Minnesota's Roadside Architecture.** View photos of several larger-than-life statues.
- ✎ Choose a statue to research on the Internet. Try to find its location. Explain the links in your search.

Minnesota

Geography

Web site: http://www.dnr.state.mn.us/ecs/ecs.htm

- Read information about Minnesota's three **Ecological Regions.**
- ✎ Draw a state map and mark the regions. Design icons to indicate each region.

Web site: http://minn.yahoo.com/

- Click on **Maps and Views.** What highway runs east-west through Minneapolis? In what part of St. Paul is the Minnesota State Fairgrounds located?

Web site: http://www.city.net/countries/united_states/minnesota/

- ✎ Use the information to make a bar graph showing the high and low temperatures in Alexandria, Duluth, Minneapolis, and Rochester. If possible, include both Fahrenheit and Celsius readings.

Web site: http://deckernet.com/minn/superior

- What is the length and width of Lake Superior? What is its depth? What is the general topography along the north shore of Lake Superior?
- What geologic events resulted in the formation of the lake? Look at the images of Lake Superior.

Cut along dashed line.

Mississippi

General Information

Web site: http://valuecom.com/mississippi/info.htm

- Search the site to find the meaning of the word *Mississippi.* What is the state capital? How many counties make up the state? What insect damaged Mississippi's cotton crop? What are the state's major agricultural crops? What goods are manufactured there?
- ✎ Do research to learn more about the boll weevil. Draw a diagram of the insect and explain how it damages crops.
- ✎ Explain Mississippi's position on slavery during the Civil War.

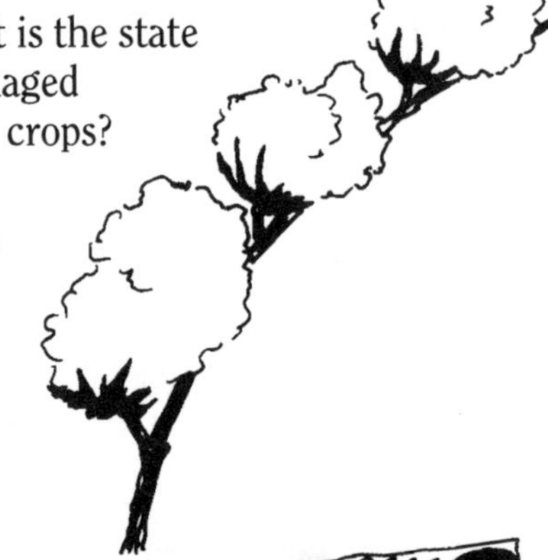

Web site: http://www.mississippi-river.com/

- Click on the **Great River Road Map** and list the ten states that border the Mississippi River. What Canadian province borders the river?
- ✎ Copy and color this map. Label each state capital.
- ✎ Search the Internet for statistics about the Mississippi River. How does it affect the economies of states along its banks?

Search words:
Jefferson Davis, Confederate States, Natchez Trace, Vicksburg, Gulf Coast

Mississippi

Cities

Web site: http://www.visitjackson.com/

- Select **City Overview,** then **Maps.** Click on the **Downtown Map** and list the nearest cross streets for any ten landmarks listed in the map key.
- Read about the **History** of Jackson. Explain its beginnings as a French trading post. For whom is the city named?

Web site: http://www.vicksburg.org/

- Click on **Convention and Visitors' Bureau,** then **Museums.**
- ✎ Do research to learn about the siege of Vicksburg during the Civil War. At what museums can you see exhibits and artifacts related to the battle?

Web site: http://www.gulfcoast.org/

- Click on **Mississippi Gulf Coast Cities,** then **Biloxi.** Locate Biloxi on a map of the state.
- Click **Visitor Info** and read **About Our Area.** Make a list of the attractions. Read **About Our Beach.** What is unusual about Biloxi's twenty-six miles of white sand beach?

Cut along dashed line.

Mississippi

Tourist Attractions

Web site: http://www.decd.state.ms.us/tourism.htm

- Go to **Itineraries** and select one that interests you. Click on all the cities included in the itinerary.
- ✎ Write a daily travel diary explaining where you went and what you experienced in each location.
- Select **Literature** and fill out a form to order the free travel brochure of your choice.

Web site: http://www.deltaqueen.com/

- Go to the bottom of the home page, click on **Next,** and read about the **History** of the paddlewheel steamboats that cruised the Mississippi River between 1811 and the turn of the century.
- Go to the bottom of the page and explain what a tourist can expect on a steamboat cruise. If you wish, order a free brochure for more information.
- Click on **Mississippi Queen.** What are the dimensions and capacity of the boat?
- ✎ What part did the steamboats play in the development of the American frontier?

Mississippi

Geography

Web site: http://valuecom.com/mississippi/info.htm

- Search the site to discover what states border Mississippi. What river forms the western boundary? What are the average temperatures in January and July? What animals live in the forests of Mississippi?
- ✎ Design a map that shows the **physical geography** of the state. Label the Mississippi River.

Web site: http://www.vicksburg.org/

- Select **Other,** then go to **Major Routes to Vicksburg.** Study the map.
- ✎ Explain in words the route from New Orleans to Vicksburg. What highways and directions would you travel from Layfayette to Vicksburg by way of Alexandria and Natchez?

Web site: http://www.hattiesburgms.com/statmap.html

- Look at the state map. Explain why Hattiesburg is called the Hub City. Where do U.S. Highways 49, 59, and 98 lead (all directions)?

Missouri

General Information

Web site: http://www.yahoo.com/Regional/U_S__States/Missouri/

- Select the **Unofficial Missouri Page,** then go to **Facts.** What are the ten largest cities?
- Research information about the state bird, flower, and tree. Draw a picture and write a paragraph about each one.
- Go to **History.** Make a time line illustrating six important events in Missouri's history.
- Who was Dred Scott? What is his connection to Missouri's history?

Web site: http://www.ecodev.state.mo.us/tourism/facts/default.htm

- Read the list of **Famous Missourians.** Write the list and use one word to describe each person's occupation.
- Which **Famous Missourian** became president of the United States? What was his most important decision while in office?
- What was Mark Twain's real name? In which city was he born? Name two characters in his popular books.
- Go to **Agriculture.** What is Missouri's largest cash crop? What other crops are produced there? What livestock is raised?

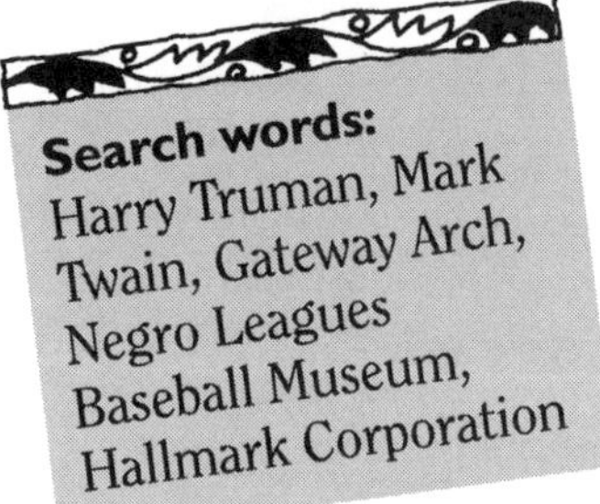

Search words:
Harry Truman, Mark Twain, Gateway Arch, Negro Leagues Baseball Museum, Hallmark Corporation

Missouri

Cities

Web site: http://stlouis.missouri.org/

- Choose **About St. Louis,** then the **Gateway Arch.** When was the arch built? Whom does it honor? What museum lies at the base of the arch? What can be seen in that museum?
- Read the **Brief History** of St. Louis. In what year and by whom was the city founded? In what year did the city host the world's fair? What is the **population** of the city? What is the racial mix according to the 1990 census?

Web site: http://webcrafters.com/kcmetro/

- Go to **KC Sights and Places,** then **Points of Interest.** Check out **Sports Attractions** in Kansas City. Name the professional baseball and football teams. How do professional sports teams contribute to the quality of life in a city?

Web site: http://www.kansascity.com/zoo/

- Visit the **New Zoo** from this site. Read **Zoo History** and summarize how the zoo has grown over the years. Look at the **Zoo Gallery** photos. Make an alphabetical list of animal names.

Cut along dashed line.

Missouri

Tourist Attractions

Web site: **http://www.mobot.org/welcome.html**

- What is the location of the **Missouri Botanical Garden?** How large is it? What does it include?
- View a wide variety of flowers in the **Image Galleries** and **Plants in Bloom.**

Web site: **http://www.ecodev.state.mo.us/tourism/**

- Look at the photos in **Tour Missouri.** Which tour would you prefer to take? Explain.
- ✎ What is Six Flags Over Mid-America? Search the Internet for information about the attractions offered there.
- ✎ Search the Internet for information about the Negro Leagues Baseball Museum. List five players who are honored there.

Web sites: **http://www.ecodev.state.mo.us/tourism/ozark.htm**
http://www.bransonconnection.com/

- What tourist attractions are located in the Ozark Mountain Region? How many days would you need to see the area? What kinds of clothing would you take along?
- What crafts are showcased at **Silver Dollar City?**

Missouri

Geography

Web site: **http://www.mobot.org/welcome.html**

- Read about the **Plant and Animal Communities** in the Shaw arboretum. Explain how the location of the Arboretum contributes to its plant and animal diversity.
- ✎ Make a chart showing the ecological communities in the arboretum and the plant and animal species that live there.

Web site: **http://guide2america.com/missouri/**

- Click on **Missouri Maps** and print the maps of north and south Missouri. Line up the maps at Jefferson City and tape or glue them together.
- ✎ Use a highlighter to mark Jefferson City, St. Louis, Independence, and Springfield on the map. Which direction would you travel from Jefferson City to each of the other cities?

Web site: **http://www.nps.gov/ozar/**

- Visit the site of the **Ozark Scenic Riverways** in southeastern Missouri. Click on the **caves** link. How many caves are located in the area? What is their ecological value?
- ✎ What are the characteristics of karst topography?

Cut along dashed line.

Montana

General Information

Web site: http://infoplease.lycos.com/states.html

- Click on **Montana** to answer the following questions. What states and Canadian provinces border Montana? When did Montana become a state? What does the word *Montana* mean? What is the state's nickname?
- ✎ What is the population density of Montana? How is that similar or different to your home state? How would the number of people per square mile affect the quality of life in Montana?
- What American Indian groups live on reservations in Montana?
- What are Montana's mineral resources?
- What is the Continental Divide? How does it affect Montana's climate?

Search words: Continental Divide, Glacier National Park, General George Custer, Lewis and Clark Expedition

Montana

Cities

Web site: http://www.remaxofhelena.com/helena.phtml

- Where is Helena located? What highways meet in the city? What national parks are nearby? Describe weather conditions in Helena.
- ✎ Use the **weather** link to check Helena's five-day forecast. Type in the names of three other cities in Montana and use the data to make a general statement about weather conditions in the state for one week.

Web site: http://usacitylink.com//billings/do.html

- What is the historical significance of the Battle of Little Big Horn? Who were the participants? What was the outcome? What national monument marks the battle?
- How is the Lewis and Clark Expedition remembered at Pompeys Pillar?

Web site: http://www.onroute.com/destinations/missoula.html

- Name four rivers near Missoula. What recreation is available on the rivers? Name the two museums in the city.
- ✎ What are bison? Where is the National Bison Range? If you visited there, what would you expect to see and do?

Cut along dashed line.

Montana

Tourist Attractions

Web site: http://www.nps.gov/glac/home.htm

- Select **Attractions** and go to **Glacier National Park.** Describe the natural beauty of the park for the current season. What species of animals live in the park? What nearby communities provide accommodations for visitors?
- Click on **Glacier National Park Website Visitor Center.** From here you can access park photos and maps. What do you think is the most interesting feature of Glacier National Park?
- Go to **Camping and Lodging,** then **Ranches and Resorts.** Read several sites. Would you prefer to stay at a resort or a ranch? What activities and accommodations would be available at each? Explain your choice.

Montana

Geography

Web site: http://www.montanawildlife.com/

- Many species of threatened and endangered **Wildlife** live in the forests of Montana. How are **Montana's Wildlife Management Areas** helping to maintain the animal population?
- ✎ Select one animal species. Click on its link and summarize the information in a brief report.
- ✎ Select an area on the map. Click on the name and record information about the park's size and location. What animal species will a visitor see?

Web site: http://www.nps.gov/glac/home.htm

- Select **Attractions** and go to **Glacier National Park.** Read the **geologic history** of the park. What is a glacier? How do they sculpt the land?
- ✎ Make an illustrated chart or pictionary of these terms as they refer to glaciers: horn, cirque, arete, hanging valley, moraine.

Web site: http://www.lib.utexas.edu/Libs/PCL/Map_collection/states/Montana.gif

- Print the GIF maps of **Montana counties** and **Indian reservations.** Compare the two maps. What counties have reservations? Print the **Largest City Map** and highlight the ten largest cities in the state.

Nebraska

General Information

Web sites: **http://www.neupc.org/NebraskaInfo.html**
http://www.ded.state.ne.us/tourism/report/symbols.html

- Click on **General Information.** What is the origin of the word *Nebraska?* What is the state capital? nickname?
- Go to the second site. What is the state tree? fossil? insect? grass? bird?
- ✎ Nebraska has a state poet. Who is he? Draw a picture to illustrate his vision of life in Nebraska.
- ✎ Draw the state seal. Explain the meaning of each of the symbols.

Web site: **http://www.state.ne.us/gov/gov.html**

- Who is the current governor of Nebraska? Read his **Biography.** What is his political party? What are his goals for improving life for the people of Nebraska?

Search words:
Chimney Rock, Oregon Trail, Mormon Trail, Cornhuskers, Platte River

Nebraska

Cities

Web site: **http://db.4w.com/lincolntour/**

- Take the **Picture Tour** of Lincoln. You may look at individual photos or select the **Locator Map** for a self-guided tour beginning in front of the capitol building.

Web site: **http://www.surflincoln.com/StuffToDo.html**

- Select **Show All Listings** of **Stuff To Do** in Lincoln. Make a list of three each: indoor activities, outdoor activities, and restaurants.

Web site: **http://omahafreenet.org/metro/**

- From this site you can access links to Omaha's history, government, attractions, and much more.
- ✎ Go to **History.** Use the information to create a time line.

Web site: **http://www.ci.scottsbluff.ne.us/**

- Click on **Community.** Where is Scottsbluff located? How many miles is Scottsbluff from Omaha? from Lincoln?
- What Natural Resources can be found in the Scottsbluff area?
- ✎ Who was Hiram Scott? How did the town develop after 1900?

Cut along dashed line.

Cut along dashed line.

Nebraska

Tourist Attractions

Web site: http://www.visitomaha.com/

- This is the site of the Omaha Convention and Visitors Bureau. You can access a listing of **Sights and Attractions.** What cultures are represented in Omaha's **museums?**

Web site: http://www.ded.state.ne.us/tourism.html

- Click on **Photo Gallery** for a look around Nebraska, then go to **Attractions.** What can you find to do in and around Bayard? Go to the **Oregon Trail Wagon Train** link. What would modern travelers like or dislike most about a trip of this kind? Explain.

Web site: http://www.ded.state.ne.us/tourism/mormon/mormon.htm

- Read the story of **Nebraska's Mormon Trail.** Who were the Mormons? Who was their leader? What was their final destination?
- Look at the **Map** of their route. Which direction did they travel across Nebraska? Read the information about the **Sites** on the current trail. Highlight five of them on a map.

Nebraska

Geography

Web site: http://www.ded.state.ne.us/tourism/region.html

✎ Draw a map of Nebraska, labeling the border states. Go to all eight links for the tourist regions. Read the information and label the regions on your map.

Web site: http://www.knb.org/

- Select **Interactive** from the choices at this site. Take the **Keep Nebraska Beautiful Trivia Challenge** (environmental quiz) and check your answers.

Web site: http://www.necga.org/

- This is the site of the National Corn Growers Association. Select **Corn Acres Harvested by State.** Look at the map and read the chart. What states harvest more corn than Nebraska? How many thousands of acres of corn did Nebraska harvest according to this data?

Web sites: http://www.prairieweb.com/scb_gering_ucc/a_chrock.htm
http://www.orecity.k12.or.us/OregonTrail/ChimneyRock

- Visit the **Chimney Rock Historical Site** and read the diary of a traveler on the Oregon Trail.

✎ Write a diary entry describing your impressions of the rock as if you are a traveler on the Oregon Trail.

Nevada

General Information

Web site: http://www.state.nv.us/mansion/symbols.html

- ✎ From this site you can learn about **Nevada's State Symbols.** Print ten of them and create a Nevada state symbols poster.
- ✎ Work with a partner to think of three new (original) state symbols that would be appropriate for Nevada. Write two reasons that support each selection.

Web site: http://www.vcnevada.com

- Read about the importance of **Virginia City,** Nevada, in the history of the American West. Go to **The Past** and read the **History of the Comstock Lode.** What was the Comstock Lode? Where was it discovered and for what was it used? What favor did Abraham Lincoln grant Nevada in 1864?

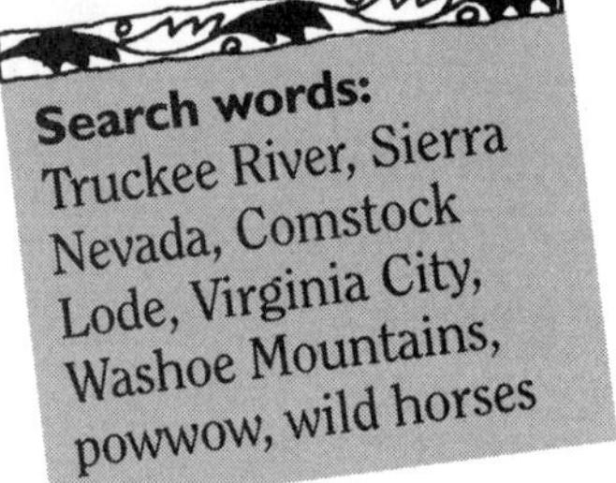

Search words: Truckee River, Sierra Nevada, Comstock Lode, Virginia City, Washoe Mountains, powwow, wild horses

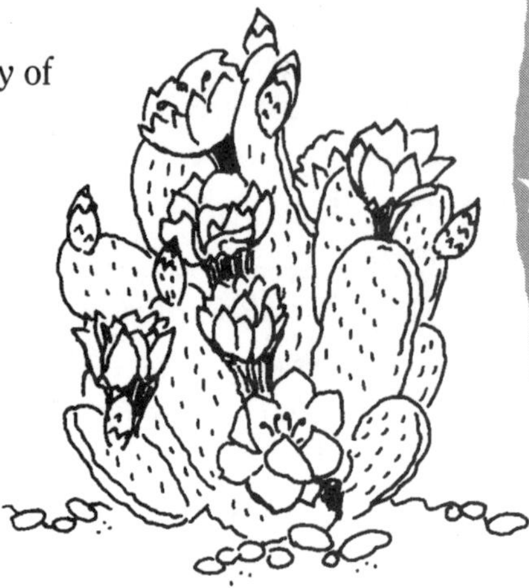

Nevada

Cities

Web site: http://www.reno.gov/

- What is the general climate of Reno? Check the current **weather conditions.** Go to the end of the page and check **Reno Events.** What is the next city event listed?

Web site: http://www.nevadaweb.com/cnt/r-t/reno.html

- Read the **History** of Reno. What was Lake's Crossing? When did the railroad reach Reno? For whom is the city named?

Web site: http://city.net/countries/united_states/nevada/las_vegas/

- Select **Factsheet.** How many people live in Las Vegas? Read the average temperatures. Which months would you need to wear a jacket or coat? When are the casinos closed?
- Go to **Sightseeing.** What is there to do in Las Vegas away from the "Strip"? Choose two museums and one outdoor activity that interest you. How much would it cost to visit all three?
- ✎ Read the entire sightseeing list. Draw a Venn diagram. Label these sections: families with children, couples over forty years of age, and single persons. List five attractions in each section.

Cut along dashed line.

Cut along dashed line.

Nevada

Tourist Attractions

Web site: **http://www.hooverdam.com/**

- Tour the **Hoover Dam Picture Gallery.** Read the **Story of Hoover Dam** and the **History of Hoover Dam.** Why was it necessary to build the dam? For whom was the dam named?
- ✎ Read the journals of workers who built the dam. What problems do they remember?

Web sites: **http://www.virtualtahoe.com/**
http://www.travelnevada.com/~ncot/renotahoe.html

- Make a list of all the types of recreation offered at Lake Tahoe. Check the **ski report** and watch live pictures on the **Tahoe Sky Cam.**
- ✎ What Nevada cities are nearest Lake Tahoe?

Web site: **http://www.travelnevada.com/**

- Visit the **five territories** of Nevada. Name two cities that are located in each one. What kinds of activities took place in the territories early in the state's history? Describe each territory's natural environment.
- ✎ Read all the information in **Cowboy Country.** Pretend you visited there. On returning home, write a short poem about your experiences.
- ✎ Design an icon for each territory and use it on your own state map.

Nevada

Geography

Web site: **http://www.geocities.com/Yosemite/Trails/8087/**

- Enjoy a hike along the **Ruby Crest Trail.** Where is the trail located? Considering the length of this hike and Nevada's climate, what would you wear and carry with you on this hike? What would you probably have the opportunity to photograph?

Web site: **http://www.nps.gov/grba/**

- Discuss the environment of **Great Basin National Park.** Where is it located?
- ✎ Go to **Directions.** Locate the park on a map of Nevada and trace the routes from the park to Reno and Las Vegas. Indicate the mileage.
- ✎ Print the **Map of the Park.** Highlight four mountain peaks and four creeks shown on the map.

Web site: **http://nimbo.wrh.noaa.gov/cnrfc/linksto.htm**

- Go to the **Weather Forecast Offices Map.** Visit the Las Vegas site and check the radar map. What are current weather conditions in Las Vegas? Explain why its weather makes Las Vegas a popular tourist destination.

Cut along dashed line.

New Hampshire

General Information

Web sites: **http://www.state.nh.us/nhinfo/nhinfo.html**
http://www.state.nh.us/senate/senkids.htm

- Welcome to the **New Hampshire Almanac.** At this site you will find many easy-to-read pages with general information about the state. Go to **History.** Why was the state founded? What was its first name? Who were the Masonian Proprietors?
- Explain New Hampshire's geographical location. Name five New England rivers that start in New Hampshire.
- What was **Fast Day?** How was it observed? When was it abolished?
- Who was the first **Governor** of New Hampshire? Who is the current governor?
- What is the official state amphibian? Why was it chosen?
- Describe New Hampshire's **state tartan.** What is the significance of each color?

Web site: **http://www.nh.com/factsfun/lovenh.html**

- Read **Five Reasons to Love New Hampshire.** Think about the importance of each item. Are the reasons good enough to convince you to move to New Hampshire?
- ✎ Search other sites about New Hampshire and write down five other reasons why it is a good place to live, work, and raise a family.

New Hampshire

Cities

Web sites: **http://www.state.nh.us/localgovt/cities.htm**
http://www.state.nh.us/soiccnh/idxcompr.htm

- Choose six to eight communities that appear on both sites and complete one of the projects below.
- ✎ Round the numbers to the nearest inch and make a bar graph showing the high and low average temperatures.
- ✎ Locate the counties on a map. Write in the community names.
- ✎ List the community populations and areas from smallest to largest, then compare the results and draw conclusions about population density.

Web site: **http://www.ci.concord.nh.us/**

- Click on **History** and select **Historical Narrative.** In what year was Concord made the seat of New Hampshire government? What U.S. president made Concord his home?
- Click **Maps** and read the **Distances to Major Cities.** What interstates cross at Concord?
- ✎ Study the map of the region. Select five cities from the chart and calculate how long it would take to reach Concord from each of them traveling an average of fifty-eight miles per hour.

Cut along dashed line.

New Hampshire

Tourist Attractions

Web site: http://www.nh.com/tourism/index.shtml

- Select **Wildlife and Woodlands.** Make a chart of New Hampshire's wildlife and fish divided into these categories: large game animals, upland birds, raptors, fish.
- ✎ What is an osprey? Do research. Write a brief report and make a drawing to share with the class.

Web site: http://www.visitnh.gov/

- What are the **six tourist regions** of New Hampshire? Explain briefly the characteristics that make each region unique.
- Study one region in depth. What attractions are located there? Write addresses for one lodging and one restaurant.
- ✎ What is Dartmouth? Name two other similar educational institutions in New England.

Web site: http://www.ci.concord.nh.us/tourdest/index.shtml

- ✎ Learn more about the Shaker way of life by visiting their site at **Canterbury Shaker Village.** Who were the Shaker people? What did they believe? How did they live? What does the village offer tourists today?

New Hampshire

Geography

Web site: http://www.nh.com/factsfun/countymp.html

- How many counties make up the state? Which is the northernmost county? Which county appears to have the longest coastline?
- ✎ Use the clickable map to find the area and population of each county. Draw a map of New Hampshire. Label the counties. Fill in their square mileage and populations.

Web site: http://www.portsmouthnh.com/

- Check the **attractions** along the beach at Portsmouth (Port City). Go to the **Photo Gallery** and view pictures of seasonal changes on the seacoast. How would weather affect the tourist trade?

Web site: http://inlet.geol.sc.edu/GRB/home.html

- Go to **Site Description** for information about the area of Great Bay Reserve. What species of flora and fauna live in the bay area? Which are endangered?
- Go to **Cultural History.** What artifacts have been unearthed in archaeological digs in the area?

New Jersey

General Information

Web site: http://www.state.nj.us/njfacts/njsymbols.htm

- Click on each picture for detailed information about each state symbol.
- ✎ Make a drawing of the state flag. How do the colors of the flag relate to the state's early settlers?
- ✎ What is the state dinosaur? Why was it chosen? Search the links for more information. Make an illustrated poster with five facts about the dinosaur.

Web site: http://168.229.3.2/NewJersey/GeneralInfo/

- Go to **General Info site B+.** What kind of **Business** is most important to New Jersey's economy?
- Go to **General Info site D+.** What are New Jersey's seven largest cities? In what year will the total population reach 8.5 million?
- Go to **General Info site H+.** Read these **Interesting Facts about New Jersey.** Explain the history of women's right to vote in the state.
- ✎ Search the Internet for biographical information about Thomas Edison. Make a list of his inventions. Chart which ones you use in a typical week.

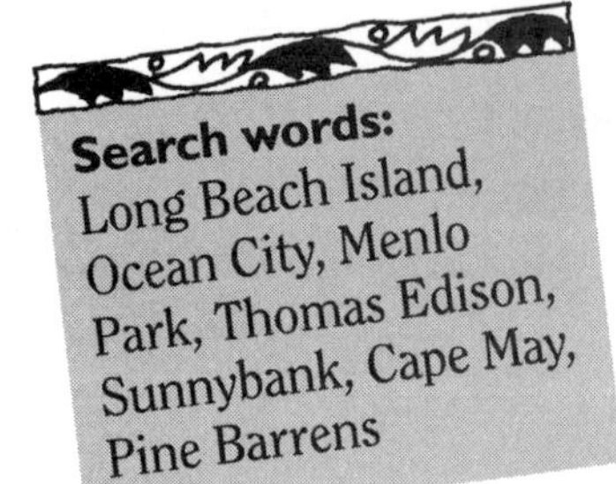

New Jersey

Cities

Web site: http://www.prodworks.com/trenton/

- This is the home page of New Jersey's capital, Trenton. Go to **Local Government,** then **City of Trenton.** Where can you contact the mayor? Go to **General Information.** In what county is Trenton located? What is the city's population?
- ✎ Search the lists of **Restaurants** and **Cultural** offerings to find a museum and a restaurant on the same street.

Web site: http://www.moon.com/road_trip/coastal_east_coast/map.html

- Study this map of the **Coastal East Coast.** Locate New Jersey on the northern coast and list the cities labeled on the map.
- Click on **Across New Jersey** to read about the cities on the coast and the attractions they offer.
- ✎ Explain the location and history of **Lucy the Elephant.** For what is it used?
- ✎ Describe how **Cape May** is different from surrounding communities on the New Jersey coast.

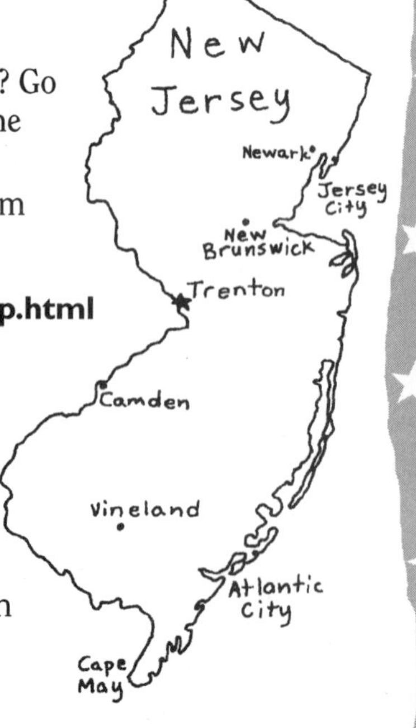

Cut along dashed line.

Cut along dashed line.

New Jersey

Tourist Attractions

Web site: http://www.atlantic-city-online.com/

- Go to **History** and select **A Look Back at Atlantic City.** What shops will you see while strolling the "boards"?

Web site: http://www.virtualac.com/

- From this site you can experience three different tours of Atlantic City. **The Virtual Boardwalk Tour** allows you to select **Go Back In Time** and view period architecture. Jitney and helicopter tours offer a broader view of the area.
- ✎ It is said that American postcards originated in Atlantic City. Study the sites and design a postcard featuring one of the attractions. Write a friend about what you saw on your virtual vacation.
- ✎ Design three souvenirs that might be purchased on the Boardwalk.

Web site: http://people.csnet.net/dpost/welcome.html

- New Jersey was the site of several battles during the American Revolution. Read **The War in New Jersey.** How did George Washington capture Trenton? What is a bayonet? Locate Trenton, Princeton, and the Delaware River on a map. In what New Jersey city did Washington and his men spend the winter?
- Click on **Historical Sites** to get an idea of what is preserved for visitors who are interested in history.

New Jersey

Geography

Web site: http://www.state.nj.us/pinelands/

- Describe the environment of **Pinelands National Reserve.** Read its **History.** How has it changed over the years? How does living in the Pinelands differ from living in urban New Jersey?
- ✎ What is an aquifer? a biosphere? Name some examples of unusual plants and animals living in the Pinelands. What crops are grown there?
- ✎ Why is maintaining the Pinelands important to New Jersey and the entire eastern seaboard?

Web site: http://woi.com/nj/weather.html

- Click on **Current Regional Temperatures Map.** Calculate the average temperature in the state or select a city and read the temperature changes over the previous twenty-four-hour period.
- ✎ Select a specific feature and time from the pull-down menu. Use the **Map Symbol Legend** to explain the new map.
- ✎ Use the **Map Symbol Legend** to create a map of your state with current conditions.

New Mexico

General Information

Web site: http://www.state.nm.us/

- Go to **New Mexico Fast Facts,** then select **Fast Facts.** When did New Mexico become a state? What is the capital city? the state's nickname?

✎ Make a *zia* (god's eye) like the one pictured on the state flag.

Web sites: http://www.caverns.com/~abctravel/nm_hist.htm
http://www.caverns.com/~abctravel/taosfoun.jpg

- These sites link with sites describing New Mexico past and present. Read about **Territorial New Mexico.** Check the links for photos. Trace the **Santa Fe Trail** on the map.
- Explain how mining boosted New Mexico's early economy. What is *bat guano?*

Search words:
Snake Petroglyph, Chaco Canyon, Bisti Badlands, Sandia Peak Tramway, Lincoln, Fort Sumner, Trinity Site

Web sites: http://listserv.american.edu/catholic/franciscan/ofm/olg/zuni.html

- Read this site about the **Zuni People** of New Mexico. How were their beliefs affected by Catholic missionaries? What crafts are they best known for? What are *kachinas?*

✎ Build a model of a pueblo. Write a story about daily life in a Zuni village.

Cut along dashed line.

New Mexico

Cities

Web site: http://www.ptig.com/sfccoc/sf_hist.htm

- Read **Santa Fe's History.** What groups of people first established a settlement on the site of Santa Fe? What is the city's full name? What is its nickname?
- What cultural institutions and outdoor activities are available in the city? Why do you think Santa Fe was named the best travel destination in the world?

Web site: http://www.roswell-online.com/

- Where is Roswell located? Why has this city become a popular tourist destination? How has life changed for local people since July 1947?

Web site: http://www.abqcvb.org/

- Read **The City's Past** and explain how Albuquerque got its name. Why has the spelling changed from the original? How did **The Coming of the Railroad** help the town grow?
- Click on **Native American Culture.** What group(s) currently have reservations near Albuquerque? What is a pueblo? What craft items are part of these American Indian cultures?

Cut along dashed line.

New Mexico

Tourist Attractions

Web site: **http://www.nps.gov/cave/**

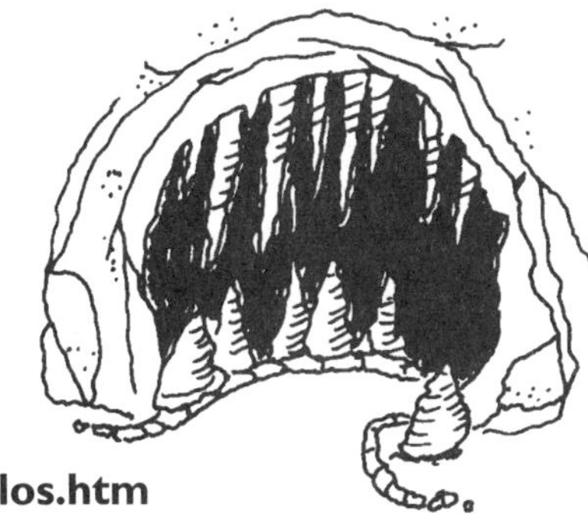

- How many caves are located in **Carlsbad Caverns National Park?** How deep is the longest one? When is the best time to visit the park? What is the cost of a guided tour of **Kings Palace?**
- ✎ Read the links that describe the more difficult cave tours. What special equipment is required for the tours? Would you enjoy this kind of experience?

Web sites: **http://www.roswell-online.com/environs/m_silos/silos.htm**
http://www.iufomrc.com/index1.html

- Take the tour of the **Missile Silos** and decide for yourself if you believe the "incident" story.
- Tour the **International UFO Museum and Research Center** in Roswell. What is the **Roswell Fragment?** Do you believe the stories about its origin?

Web site: **http://www.state.nm.us/MOIFAOnLine/**

- The **Museum of International Folk Art** is a part of this state's Office of Cultural Affairs. Look at the **permanent collections and exhibits.** What is New Mexico's cultural heritage?
- ✎ What materials were used in objects in this collection? What are *bultos? retablos?*

New Mexico

Geography

Web site: **http://www.state.nm.us/nmmag/area.html**

- Click on the areas to read detailed information about New Mexico's attractions.
- ✎ Draw a map like the one on the home page. Click on each of the areas and read the information. Label or draw natural landmarks and historic sites in each area.

Web site: **http://www.sdc.org/nmwc/blmwild.html**

- Read **BLM Wilderness in New Mexico.** What is the New Mexico Wilderness Alliance doing to help preserve the state's wilderness? Why is its work important?
- ✎ What is biodiversity? Design a diorama showing a desert ecosystem.
- ✎ Make a list of the lands proposed for wilderness designation. Explain the reasons for each recommendation.

Web site: **http://www.state.nm.us/state/FastFacts/Welcome.html**

- Read **County Maps** and **Environment and Geography.**
- ✎ Print the map and highlight these cities: Albuquerque, Santa Fe, and Silver City. Fill in the names of bordering states. Mark the highest and lowest points in New Mexico on your map.

New York

General Information

Web site: http://www.nuwebny.com/

- Read **General Information.** How did New York get its name? What is its largest city? What is the state bird? flower? tree?
- Click on **Arts and History.** Look at several cities on the clickable **State Map.**
- Go to **New York History.** How were the Adirondack Mountains and Finger Lakes formed? People from what European country first explored the state?

Web site: http://www.theinsider.com/nyc/index.html

- This site has information to prepare you for a trip to New York City. Go to the **Top 5 Places To Take Kids.** Visit all five links and explain what is available for children.

Web site: http://www.neinfo.net/~Fort_Ticonderoga/

- Read the **American Revolution** portion of the **Timeline** and view **Geographical Drawings** of the fort. What American captured Ticonderoga from General Burgoyne and the British?
- Go to **Photo Gallery** and view the **Scenes of the Fort and Encampments.**

✎ Write a diary entry for one of the American soldiers who helped capture Fort Ticonderoga.

Search words:
Catskill Mountains, Niagara Falls, Erie Canal, Yankee Stadium, Bronx Zoo, Manhattan, Empire State Building

New York

Cities

Web site: http://ny.yahoo.com

- This is the Yahoo! search engine for New York City. From this site you can access **live views** and **photos** of sights around the city. Read current statistics and schedules for professional sports teams.

✎ Choose one sport (football, baseball, or hockey). Search the Internet to learn the names and positions of players on New York City's professional team.

Web site: http://www.albanyny.org/

- How far is **Albany** from New York City? What three highways lead to Albany?
- Click on **Visit Our Past.** When did Albany become the state capital? Who was Henry Hudson? How did he contribute to the early settlement of Albany?

Web site: http://www.our-hometown.com/

- Here you can visit the home pages of a hundred communities in western New York. Read the information about any six communities.

✎ Select three communities. Compare them on a Venn diagram.

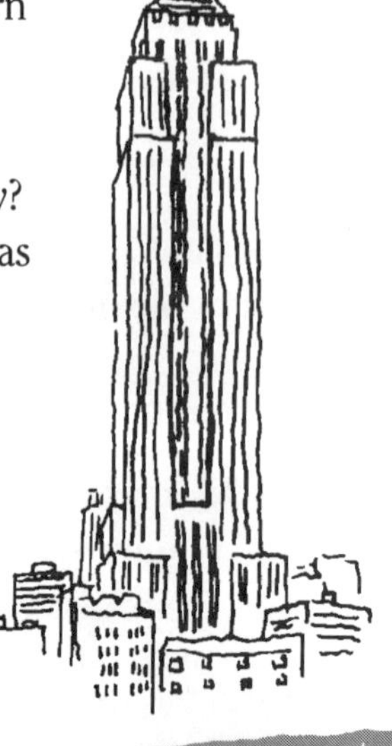

Cut along dashed line.

Cut along dashed line.

New York

Tourist Attractions

Web site: http://www.iloveny.state.ny.us/attractions/

- Use the clickable map or the list to choose an attraction that interests you.
- Where were the 1980 winter Olympics held? What events are held in the area? What is Whiteface? In what city is the **National Baseball Hall of Fame and Museum? Belmont Park Race Track? Carnegie Hall?**

Web site: http://www.nps.gov/stli/

- Click on **Expanded Homepage.** Read the information about the history of the Statue of Liberty and Ellis Island. On what date was the statue dedicated? Who was its sculptor? What is inscribed on the statue's tablet?
- ✎ Study the **Statistics of the Statue.** Make life-sized eyes, a nose, and a mouth from paper and assemble her face.

Web sites: http://www.centralpark.org/home.html
http://www.theinsider.com/nyc/attractions/2central.htm

- Click on **Find It** and view a map of **Central Park.** What streets form the east and west boundaries? At what cross streets will you find the following: Tavern on the Green, the band shell, the Metropolitan Museum of Art, and the police department?

New York

Geography

Web sites: http://www.nfcvb.com/
http://niagara.buffnet.net/Photos/

- Click on **Falls Facts.** Name the three falls. How old do geologists estimate the falls to be? At what rate does water flow in the Niagara River? For what is that water power used?
- View the photos of Niagara Falls. Go to **Geography.** What is unusual about the location of Niagara Falls? Which Great Lakes are located nearby? What cities are connected by the shipping routes of the Niagara Falls region?

Web site: http://www.mediabridge.com/nyc/histfacts/geography.html

- What is the geographical location of **New York City?** Read the information about New York City's boroughs.
- ✎ Draw a map with the boroughs labeled. Summarize facts from your reading directly on the map.

Web site: http://www.fingerlakes.net/

- Look at the map of the **Finger Lakes Region.** In which direction must you travel to reach Buffalo? Albany? New York City? What **State Parks** are located in the region?
- ✎ Print the map. Design and add icons to indicate if the park offers boating, fishing, or swimming.

North Carolina

General Information

Web site: http://hal.dcr.state.nc.us/NC/GEO/GEO.HTM

- What is the population of **North Carolina?** What is the state's total area? Explain what products are grown or produced in each of the three topographical regions.
- ✎ What feature of the state's topography is most important to the tourist industry? Explain.

Web site: http://hal.dcr.state.nc.us/NC/HISTORY/HISTORY.HTM

- What two cities have been the state capital? Name three U.S. presidents who were North Carolina natives. When and why did North Carolina secede from the Union?
- ✎ What event took place at Kill Devil Hill in 1903? How did it affect the future of world travel? Search for information and tell the story of the inventors.

Web site: http://hal.dcr.state.nc.us/NC/SYMBOLS/SYMBOLS.HTM

- Explain how the emerald **(state stone)** and sweet potato **(state vegetable)** achieved their status.
- ✎ How did North Carolina get the nickname **Tar Heel State?**

Search words:
Outer Banks, Kitty Hawk, Wright Brothers, Biltmore Estate, Smoky Mountains

Cut along dashed line.

North Carolina

Cities

Web site: http://www.virtualraleigh.com/

- Go to **General Information.** What major cities make up the Triangle Area? Raleigh is the capital of North Carolina. It has been recognized nationally as a very good place to live and raise children. Explain why.
- ✎ Read about Raleigh's accomplishments. Choose one and design a certificate or trophy to honor it.
- What **leisure** activities are popular with residents?
- Take a virtual tour of the **State Capitol Building.**

Web site: http://charlottecvb.org/ncnw/clt/

- Use the North Carolina **map** to locate Charlotte. Read **Things to Do** and **Places to See.** What could you see and do in North Carolina that you could not see or do in any other state?

Web site: http://ncnet.com/ncnw/nc-map-c.html

- Use the clickable map to access information about several cities in North Carolina.
- ✎ Choose a site to read in detail and create a packet of information that could be given to a family moving into a new community.

Cut along dashed line.

North Carolina

Tourist Attractions

Web site: http://www.nctraveler.com/

- Go to **Places.** What sights of western North Carolina attract visitors? Name the American Indian reservation and the amusement park in this region. Explain their locations.

Web site: http://www.biltmore.com/

- The **Biltmore Estate** in Asheville, North Carolina, is one of America's finest castles. Read the **History** of the mansion. Go to **Construction.** How many rooms are in the house? Who were the architect and landscape designer? Take tours of the **house** and **garden.**

Web site: http://great-smokies.com/

- From this site you can view tourist attractions in the **Smoky Mountains.** The **Museums** exhibit the crafts of the Appalachian people and American Indian groups living in the region.
- ✎ Review the information in the links and look at the pictures. What would you expect to see on a **scenic drive** at dawn or dusk? at noon?
- ✎ What artifacts would you see on display at the Oconaluftee Indian Village, Museum of the Cherokee Indian, and the Cherokee Heritage Museum and Gallery? Search the Internet for photos of Cherokee crafts or artifacts.

North Carolina

Geography

Web sites: http://www.nps.gov/caha/
http://www.beachsurf.com/srfsound/activites.html
http://www.outer-banks.com/lighthouses.html

- **Cape Hatteras National Seashore** stretches over seventy miles of barrier islands. What activities draw tourists to the island?
- View three photos of the **Cape Hatteras Lighthouse.** Why do you think the cape has been known as the Graveyard of the Atlantic?

Web site: http://www.webcom.com/cccom3/ccvb/ncmap.gif

- Study the map of **North Carolina.** Name six states that border it.
- What highway(s) would you take and which direction would you drive from Charlotte to Chapel Hill? from Fayetteville to Raleigh?

Web site: http://thesmokies.com/

- Click on **Meterological Observations.** What is the relationship of station elevation to high temperature and snow depth?
- ✎ Make a line graph showing the relationship of high temperature to elevation at all five weather stations.

North Dakota

General Information

Web site: http://www.state.nd.us/www/demographics.html

- From this site you can access information about North Dakota's **state symbols and emblems.** What is the **Honorary Equine?** Why was it chosen? Where do these horses live today?
- ✎ Read about the state seal and make a sketch. Compare your drawing with the actual seal.

Web site: http://www.ndtourism.com/info.html

- Click on **Facts.** What are the average high and low **temperatures** in North Dakota? What is the **population** per square mile? What are the **largest cities?** What agricultural products come from the state?
- Learn about North Dakota's **Historical Highlights.** Who were the first inhabitants? What did they call their homes?
- Who was the first non-American Indian visitor? What explorers arrived in 1804? What European immigrants settled the Dakota territory?
- Go to **Native Americans.** Name the individual Sioux nations. What beliefs do they share? What is the significance of a Sioux powwow?

Web site: http://www.ndtourism.com/quiz.html

- ✎ Take the **Trivia Quiz.** Check your answers.

Search words: Fort Totten, Turtle Mountain, Sitting Bull, International Peace Garden, Sioux Indians

Cut along dashed line.

North Dakota

Cities

Web site: http://www.und.nodak.edu/gifs/gf.html

- Enlarge the **Photos of Grand Forks,** then go to **Information.** What is Grand Forks' sister city? Read the **Area Attractions** and create a phone directory that includes a golf course, two museums, a speedway, a theater, and "Jugville."

Web sites: http://www.hotwired.com/rough/usa/great.plains/nd/cities.html

- How did the city of Bismarck get its name? What is the geographical location of Bismarck? How many people live there?
- Read the information about **Crime.** Why do you suppose North Dakota has such a low crime rate? What services (other than arresting criminals) could be provided by the police force?

Web site: http://www.webcom.com/nddirt/cit/cit.html

- This interesting site shows locations in North Dakota that have the same names as famous cities around the world.
- ✎ List the names. Indicate the countries where the foreign cities with those names are located.

North Dakota

Tourist Attractions

Web site: http://www.ndtourism.com/

- Click on **Activities** for a guide to recreation and sports organized by region.
- Select **Attractions** and read about the various North Dakota museums, landmarks, and state parks.
- ✎ Where is the **International Peace Garden?** What pledge has been made between the bordering nations? What regulations are enforced at the Canadian border?
- ✎ What did **Lewis and Clark** find as they explored the northern part of the Louisiana Purchase? How were they received by the American Indians? What sites in North Dakota mark the route of Lewis and Clark?

Web site: http://www.glness.com/motorcoach/attractions.html

- This site offers a detailed list of **attractions** around the state with a brief explanation of each. Some of them are unusual. Where is the world's largest buffalo? the geographical center of North America? a life-sized triceratops and pachycephalosaurs?

Cut along dashed line.

North Dakota

Geography

Web site: http://www.ndtourism.com/Geography.html

- What are the geologic regions of North Dakota? How was each one formed? How is each region used today?
- What are the highest and lowest points in the state?
- ✎ Draw a map of North Dakota. Label the highest point, the lowest point, and the geologic regions.

Web site: http://www.npsc.nbs.gov/

- Read the **Feature of the Month** at the site of the **Northern Prairie Research Center.** Explain the current topic and view the graphics.

Web site: http://www.ndsu.nodak.edu/instruct/schwert/nd_geol/nd_index.htm

- Read **Glacier Features in North Dakota.** Click on each of the photos and read the description of each area.
- ✎ Print a North Dakota county map and label each glacial feature.

Ohio

General Information

Web site: http://www.state.ohio.us/

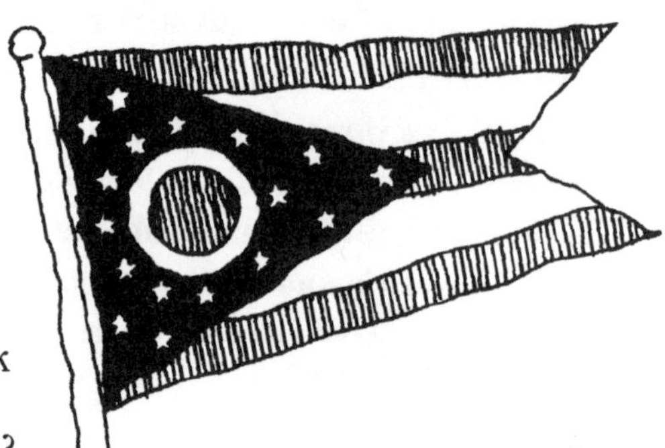

- Who is Ohio's governor? Go to the **Governor's Home Page** and read about Ohio's **history.** What is the meaning of the word *Ohio?* What American Indian tribes lived in Ohio? Where was the first permanent settlement in Ohio after the Revolution? How many counties make up the state?
- Go to **Facts About Ohio.** Why does Ohio have a state rock song? What is Ohio's state beverage? When did Ohio become a state? Which U.S. presidents were born in Ohio? What is unusual about the shape of Ohio's state flag?
- Read **Geography.** What is the state capital? the largest county? the northernmost and southernmost points?
- ✎ Click on **Science and Flight.** Make a book or a display showing Ohio's connections to famous scientists, inventors, and astronauts.

Search words:
Neil Armstrong, German Village, Ohio Village, Ohio State University, Ohio River

Cut along dashed line.

Ohio

Cities

Web site: http://dir.yahoo.com/Regional/U_S__States/Ohio/Cities/Columbus/
http://www.ohiohistory.org/

- List the **Historic Sites** in Columbus. What major university is located in the capital? Click on **Museums and Galleries.** What can you see at the Ohio Historical Center? What is the *Santa Maria?*

Web sites: http://www.state.oh.us/statehouse/welcome.html
http://www.statehouse.state.oh.us/REV.HTML

- Take a tour of **Ohio's statehouse.** What are the **rotunda** and **atrium?** Use the information to draw a sketch map of the capitol square complex. Why was the capitol renovated?

Web site: http://www.cincy.com/contents

- Click on **Convention and Visitors Bureau.** What geographic factors limit the growth of the downtown area? Look at the **Scenic Views.**
- Select the **Cincinnati Zoo** and visit the **Collection of Plants and Animals.** Explain the zoo's conservation programs.
- ✎ Make a poster describing five things that people can do to help save the environment for the world's animal population.
- ✎ Go to **Geography.** Where is Cincinnati's Airport? Print a map and trace the route from downtown Cincinnati to the airport.

Cut along dashed line.

Ohio

Tourist Attractions

Web site: http://www.inet-ohio.com/points.htm

- From this site you may access Ohio's many amusement parks, zoos, museums, and regional attractions. Choose one city and visit points of interest there.
- Be sure to check **Inventure Place** (Akron), **National Inventors Hall of Fame** (Akron), **Pro Football Hall of Fame** (Canton), and the **Rock & Roll Hall of Fame** (Cleveland).
- ✎ What is a Hall of Fame? Think of a category that interests you and design a Hall of Fame that includes at least ten individuals.

Web site: http://www.hockinghills.com/

- In what part of the state is **Hocking Hills?** Click on **Maps.** What route would you take from Columbus to Hocking Hills? If you continued south on that route, what city would you reach?
- Go to **Photo Gallery** to see the natural beauty of the region.

Web site: http://www.tw-rec-resorts.com/

- This is the site for Ohio's eight state park resorts. Use the clickable map to read about each one.
- ✎ Draw a map of Ohio showing the park resorts. Locate and label one attraction near each resort.

Ohio

Geography

Web site: http://www.columbus.net/index.html

- ✎ Print the **Map of Downtown.** Use a highlighter to locate the statehouse, bicentennial park, city center, and the bridges over the Scioto River. What state route runs north-south through the city just east of the Scioto River? Where is German Village?

Web site: http://www.dnr.ohio.gov/odnr/recycling/

- Read about Ohio's commitment to **Recycling.** What items are being recycled? Go to **Scrap Tire Recycling.** Why are tires a serious environmental problem? What can be made from reclaimed rubber?
- ✎ Select **Waste Reduction,** then **50 Ways to Reduce Waste.** Read the list. Make a list of ways to reduce waste at school.

Web sites: http://www.state.oh.us/agr/quickag.html
http://www.state.oh.us/agr/commfact.html

- Ohio's number one industry is agriculture. How many acres in Ohio are farmland? What jobs are created by the agricultural industry?
- ✎ Select five products and read their fact sheets. Rate them in order according to value. Search the Internet for a state map. Print the map and draw icons of the products in the counties where they are grown.

Oklahoma

General Information

Web sites: **http://www.otrd.state.ok.us/studentguide/**
http://www.tulsaweb.com/okinfo.htm

- Click on **History.** What was the Trail of Tears? Name the five tribes that were forced off their ancestral land.
- When did Oklahoma become a state? What natural resource made statehood a certainty?
- ✎ Read the names of famous people from Oklahoma. Choose one and write a brief biography.
- Click on **Emblems.** Explain the significance of the design on the state flag and seal. What are the state colors?
- ✎ Read **Facts.** What states border Oklahoma? What are the main agricultural products? Design a state coat of arms in green and white.

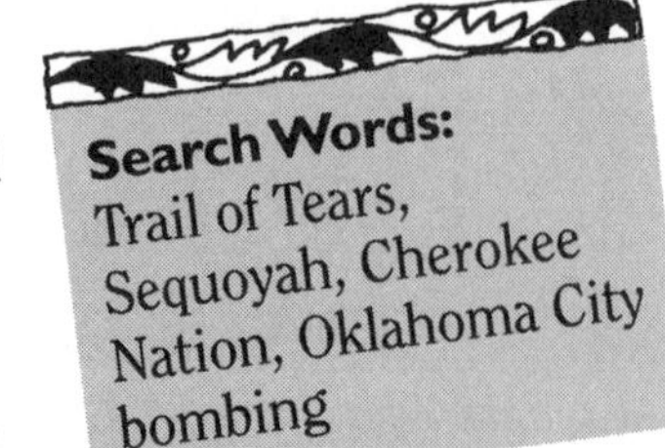

Search Words:
Trail of Tears, Sequoyah, Cherokee Nation, Oklahoma City bombing

Web sites: **http://www.adventure.com/encyclopedia/america/grapes.html**
http://www.dc.peachnet.edu/~pgore/students/w97/tanyacox/dustbowl.htm

- Read the information about the **Dust Bowl.** What states were involved? What effect did the drought have on Oklahoma's farm families? What famous book was written about this period of history?

Cut along dashed line.

Oklahoma

Cities

Web site: **http://www.oklahomacity.com/**

- Click on **General City Information.** Read about the city's historic neighborhoods and view some early photos. Go to **City Tour.** Go to **Events.** What is on the **Kids Kalendar** this month?

Web sites: **http://www.cnn.com/US/OKC/bombing.html**
http://pages.prodigy.com/GWPM57A/okctymem.htm

- Oklahoma City was the site of the worst act of terrorism ever committed on U.S. soil. Read the information and explain what happened, when it occurred, how many lives were lost, and who was responsible for the tragedy.
- ✎ What is being planned for the site of the former federal building? Draw a diagram or sketch map of the memorial.

Web site: **http://www.hud.gov/local/okl/oklacomm.html**

- This is a good site for links to many of Oklahoma's cities and towns.
- ✎ Choose a city and read the information. Write a letter to a friend as though you are a resident telling why he or she should visit.

Cut along dashed line.

Oklahoma

Tourist Attractions

Web site: http://www.okccvb.org/index.html

- From this site you can access the **National Cowboy Hall of Fame.** What artifacts are on display from frontier days?
- ✎ Make a list of things a pioneer family would have carried in a covered wagon.
- ✎ What businesses would you find in a typical Old West town? Design a diorama showing a main street or the interior of one business.
- Click on **Touring OKC** and summarize the information about these locations: Guthrie, Skirvin Hotel, Myriad Botanical Garden, Brictown, and Heritage Hills.
- ✎ What is Route 66? How is the road marked through Oklahoma? Who was Cyrus Avery? Explain the importance of the road to American culture.

Oklahoma

Geography

Web site: http://www.otrd.state.ok.us/studentguide/indians.html

- What American Indian groups have lived in the Oklahoma Territory since 1800? Explain how American Indians were relocated to accommodate westward expansion. What part did the federal government play in the conflicts between American Indians and settlers?
- ✎ How many American Indian nations live in Oklahoma today? In what ways are their cultures preserved?

Web site: http://www.deq.state.ok.us/info.html

- Read the **Fact Sheets** from Oklahoma's Department of Environmental Quality. How can you conserve water indoors and outdoors? What health problems can be caused by lead in drinking water?
- ✎ Print the fact sheet for **composting,** take it home, and begin saving organic material.

Web sites: http://cnn.com/WEATHER/NAmerica/sc/radar_image.html
http://cnn.com/WEATHER/html/OklahomaCityOK.html

- Read the temperatures and forecasts for Oklahoma City weather. Study the **radar image** for weather conditions across Oklahoma.
- ✎ Write a weather report for Oklahoma City that could be read on the evening news.

Oregon

General Information

Web site: **http://www.sos.state.or.us/BlueBook/1997_98/intro.htm**

- Go to **Symbols.** What is the origin of the name *Oregon?* Who are the "Mother" and "Father" of Oregon? Why were they selected for this honor? What is the state nut? insect? bird?
- Go to **Almanac.** What states and body of water border Oregon? How many miles is it across the country from Portland to New York City?

Web site: **http://empnet.com/mikeblee/oregon/welcome.shtml**

- You can view a **slide show** of Oregon from this site, or click on **more information** to read summaries of several categories. What is the state capital? On what date did Oregon become a state?
- What are Oregon's major agricultural and industrial products?
- Read about Oregon's **History.** Explain the importance of the Oregon Trail and railroads in the development of the territory.
- ✎ Draw a map showing the Oregon Trail from Missouri through the Northwest Passage to Oregon.

Search words:
Mt. Hood, Cascade Mountains, Crater Lake National Park, Northwest Passage, Pittock Mansion

Oregon

Cities

Web site: **http://www.ci.portland.or.us/**

- Select **General Information** and **About Portland.** Read the information to understand the factors that make Portland a successful city. Use the links to weather, neighborhoods, and events for further information.
- ✎ List five reasons Portlanders have to be proud of their city.

Web site: **http://www.efn.org/~sgazette/eugenehome.html**

- Go to **General Information** about Eugene, Oregon's second largest city. What is the population? the area in square miles? Who founded the city? What is Eugene's average yearly temperature? inches of rainfall?

Web site: **http://www.oregonlink.com/index.html**

- Welcome to **Salem Online.** Read **About Salem.** Who were the city's first settlers? What was Salem's original name? How did the Oregon Trail affect growth in the region?
- Go to **Oregon Capitol.** What is the address of the capitol? What hours is it open to the public? Look at **Views from the Observation Deck.**

Cut along dashed line.

Cut along dashed line.

Oregon

Tourist Attractions

Web site: http://www.pova.com/attract/

- From this site you can access a city map of Portland with a clickable list of **attractions.** What is the grotto? Pittock Mansion? Which museum would you be most likely to visit? Explain.

Web site: http://www.oregonlink.com/scva/historic_tour/index.html

- Use the **Historic Tour Map** to visit attractions and museums in Salem. What will you see at Mission Mill? Gilbert House? Deepwood Mansion?
- ✎ What attractions will you pass if you walk from Mission Mill to Gilbert House? In which direction would you be traveling?

Web site: http://www.llbean.com/parksearch/parks/9379GD7319GD.html

- **Mount Hood** is the highest point in the Cascade Mountains. What recreational activities are offered there?
- ✎ What is a dormant volcano? Locate and name three other dormant volcanoes in the United States.

Oregon

Geography

Web site: http://www.isu.edu/~trinmich/Oregontrail.html

- You can learn all about the **Oregon Trail** from this site. Why was it so important to the settlement of the West? What hardships and dangers did pioneers endure on the trail?
- ✎ Explain the journey of Lewis and Clark. Who sent them? What were their goals? What states were formed from the Louisiana Purchase?
- ✎ Go to **Jumping Off.** Describe the wagons used by the pioneers. Write a supply list for a family of four traveling the Oregon Trail.
- ✎ Trace the route west on a map of the United States. "Jump off" the Missouri River at Independence, Missouri.

Web site: http://www.efn.org/~sgazette/images/oregonmap.gif

- Oregon's three major cities are situated along a single highway. Describe their geographical location in the state. Which city is farthest north?

Web site: http://vulcan.wr.usgs.gov/Volcanoes/Hood/description_hood.html

- Read the information about the **Mount Hood** volcano. What type of volcano is it? How high is it? When did it last erupt? What other volcanoes are located in the **Cascade Range?**
- ✎ List all the volcanoes in the Cascade Range and indicate the state in which each one is located.

Pennsylvania

General Information

Web site: **http://www.state.pa.us/PA_History/**

- Click on **Symbols.** What is the meaning of Pennsylvania's nickname, the Keystone State?
- Read about the **State Seal.** What three symbols have also appeared on official county seals since colonial times?
- Who wrote the state song?
- Read about **Pennsylvania's Physical and Natural Properties.** What states border Pennsylvania? What are the state's three main rivers? What animals and birds live in the forests?

Web site: **http://www.state.pa.us/visit/_po-viw1.htm**

- View the beauty of the **Pocono Mountains** and discover some of Pennsylvania's connections to the past. What cities are located in the area? How did the Industrial Revolution (Lackawanna Coal Mine and Steamtown National Historic Site) affect development of the area?

Search Words: United States Brig *Niagara,* USS *Pennsylvanian,* Valley Forge, Gettysburg, Amish, Benjamin Franklin, Rodin Museum

Pennsylvania

Cities

Web sites: **http://www.nealcomm.com/pi/index.htm**
http://www.libertynet.org/mmp/

- At these sites you can take a **Virtual Tour of Philadelphia** or click on **maps** and **photos** for a closer look at the city. What natural scenery could you enjoy within a day's trip of Philadelphia?

Web site: **http://www.visithhc.com/**

- From this site you can learn about the Capital Region, which includes **Harrisburg, Hershey, and Carlisle.** Go to **Harrisburg.** Where is the city located? What cultural activities are available in the city? Request visitors' information by e-mail.
- ✎ Search the Internet for a tour of Hershey, Pennsylvania. Why is the city unique? Why is it a good place for a family vacation?

Web site: **http://www.pittsburgh.net/**

- Read the **Mayor's Welcome** to Pittsburgh. Who is the mayor? How has industry in the city changed over the years? Read the statistics that make Pittsburgh a **Livable City.** What factors contribute to a city's livability? How does Pittsburgh compare to other Pennsylvania cities in rankings?
- Go to **Visiting Pittsburgh.** What are the **Top Ten Reasons for Visiting Pittsburgh?**
- ✎ Write a list of top ten reasons to visit your hometown or a city you have visited.

Cut along dashed line.

Pennsylvania

Tourist Attractions

Web sites: http://libertynet.org/iha/valleyforge/
http://www.vaportrails.com/USA/USAFeatures/ValleyForge/ValleyForge.html

- Explain the plight of the Continental Army during the winter of 1777–78 at **Valley Forge.** Who was its leader? Was a battle fought there? What buildings still stand on the site?

Web site: http://www.gettysburg.com/battle.html

- Explain why **Gettysburg** was the most important battle of the Civil War. Detail the events on each of the three days. Who was Robert E. Lee? George E. Pickett? George G. Meade?
- ✎ Search the Internet or a reference book for the text of Abraham Lincoln's Gettysburg Address. On what date was it delivered?

Web site: http://nw3.nai.net/~spyder/

- Enjoy a visit to the **Liberty Bell Virtual Museum.** Where was the bell made? How was the bell cracked?

Web site: http://www.libertynet.org/iha/tour/_indhall.html

- What is the historical significance of **Independence Hall?**
- ✎ Search the Internet or a reference book for the names of the men who signed the Declaration of Independence.

Pennsylvania

Geography

Web site: http://www.sphpc.org/

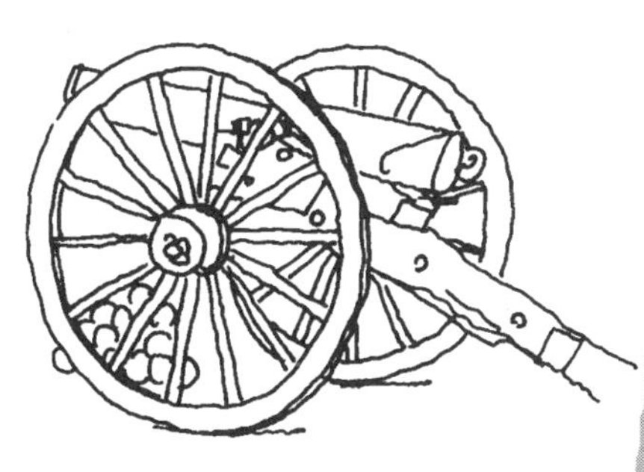

- Take a **state heritage tour** through nine counties of Pennsylvania by clicking on the **Path of Progress**. From the **Allegheny Experience Main Map,** you can access information about twenty-one historical parks, landmarks, and museums.
- ✎ Divide the map into quarters. Mark the sections northwest, northeast, southwest, and southeast. Categorize the list of attractions according to their locations on the map. What four attractions can be found near the center of the state?

Web site: http://www.800padutch.com/

- Read the **Overview of Pennsylvania Dutch Country.** Why do tourists find the area appealing? What Amish-made products can be purchased there?
- ✎ How would your life be different if you lived with the Amish? Make a drawing that illustrates the slower pace of their lives.
- Go to the **Table of Contents** and locate the **Map of Lancaster County.** How many miles is it from Philadelphia? What highways lead to Lancaster?

Cut along dashed line.

Rhode Island

General Information

Web site: http://www.state.ri.us/

- Read the **Early History** and **Independence** sections of the Rhode Island history link. What American Indian tribes lived there first? What have archaeologists been able to learn about their daily life? Why was the colony of Providence founded? Who was Roger Williams?
- Go to **Emblems.** What is the significance of the colors and symbols on the state flag?
- Click on **State House Tour.** Where is the State Room? Describe the furnishings. What is displayed in an electronically alarmed case?
- Read the **State House Statistics.** What is special about its marble dome? What other world-famous buildings have similar domes?

Web site: http://www.ritourism.com/

- Go to **More Info** and find out the population of the state. What is the capital city?
- Read **The Land.** What are the geographical regions? How did glaciers affect the terrain? What are the highest and lowest points in the state?
- ✎ Make a bar graph comparing the temperatures of Providence and Block Island for one year. Write a summary explaining the range of temperatures during the summer months.

Search words:
Roger Williams,
Narragansett Bay,
Block Island

Rhode Island

Cities

Web sites: http://www.providenceri.com/
http://usacitylink.com//providen/
http://cityguide.lycos.com/newengland/ProvidenceRIa.html

- What can you see for free in Providence? What is the Arcade?
- What is the four-day weather forecast for Providence?
- Select the **clickable map of Providence.** Choose and view specific locations. What Ivy League university is located in Providence?
- Study the map. Name four communities in greater Providence.

Web site: http://www.travelchannel.com/spot/newport/welcome.htm

- On what island is Newport located? Why was the city originally founded? Why did it become known as home to America's royalty?

Web site: http://www.visitnewport.com/

- Go to **Places to Visit,** then **Newport's Famous Mansions.** Who founded the city? How was its location a factor in its rise to prominence? Why was Newport called the Carolina Hospital?
- ✎ Do further research on the Breakers, a cottage built by Cornelius Vanderbilt. Describe the furnishings. Explain three social activities that were common at the Breakers in the 1890s.

Cut along dashed line.

Rhode Island

Tourist Attractions

Web site: http://www.ritourism.com/

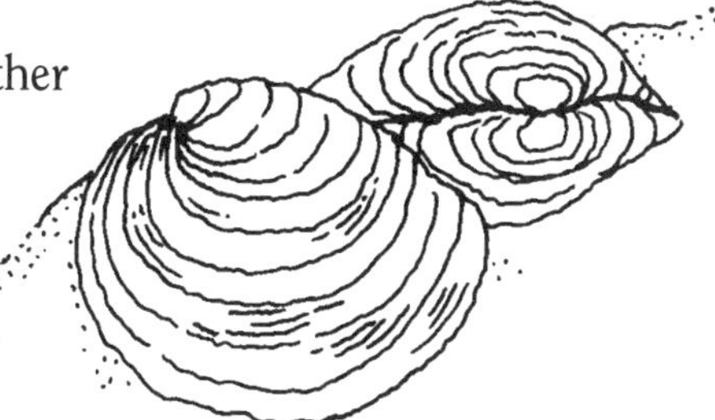

- Why is Rhode Island called the Ocean State? Name three other New England tourist destinations a day away by car from Providence. Go to **What to Do** and describe three attractions available along the coast of Rhode Island.
- Select **Block Island.** How can visitors reach the island? Why is it a popular vacation destination?
- ✎ Request that a Rhode Island vacation kit be mailed to your home.

Web site: http://www.visitri.com/places.htm

- Go to **Places to Visit.** What famous people (families) owned mansions in **Newport?**
- Describe what a visitor would see at Benefit Street's Mile of History in **Providence.**
- ✎ What is the America's Cup? Where is the America's Cup Hall of Fame? What is displayed there?

Rhode Island

Geography

Web sites: http://www.providenceri.com/narragansettbay/the_islands.html
http://www.providenceri.com/narragansettbay/the_living_bay.html

- Read the information about the **islands** in Narragansett Bay. Separate the islands into two lists: inhabited and uninhabited. Explain the topography of the uninhabited islands.
- Explain the geography of Narragansett Bay. Name the three principal islands that lie within the bay. What is the topography along the shoreline of the bay?
- ✎ Design an experiment to prove that water erodes sand, gravel, and ice.

Web site: http://zuma.lib.utk.edu/lights/ri.html

- ✎ Make a chart showing the seven Rhode Island **lighthouses** featured at this site, their locations, dates of completion, and descriptions. What is the main value of the lighthouses today? Label a map with the lighthouse names.

Web site: http://www.providence.com/directions.html

- ✎ Print a regional map. Highlight the routes to Providence.

Cut along dashed line.

South Carolina

General Information

Web site: http://www.state.sc.us/schist.html

- Read the **Brief History of South Carolina.** Who were the first European explorers in the area? What tribes of Native Americans inhabited the land? For whom was the colony named?
- Explain South Carolina's involvement in the Civil War. What was its position regarding slavery?
- ✎ What is indigo? How is it grown? For what is it used?

Web site: http://www.lpitr.state.sc.us/symbols.htm

- From this site you can see the state **seal** and **flag** and take a tour of the **State House.** Who was Major John Niernsee?
- Read **Plants and Edibles.** Why was the palmetto chosen as the state tree? Why was iced tea selected as a symbol of hospitality?
- ✎ Learn all of South Carolina's symbols and emblems. Try taking the **symbols and emblems quiz.**

Search words:
Tidewater, Myrtle Beach, Isle of Palms, palmetto, Piedmont

Web site: http://www.lpitr.state.sc.us/kids.htm

- Use this site to learn about South Carolina's general assembly. Click on **Who's Who at the State House?** and **What Are They Doing Today?** to better understand the structure of state government.

South Carolina

Cities

Web site: http://www.sciway.net/ccr_city.html

- Select **Charleston,** then **Overview.** Who is the mayor of Charleston? Read the **History** of the city. On what date was Fort Sumter attacked? What was the significance of the attack?
- Read **Geography.** What is the population of Charleston? What two rivers meet in the harbor? How does the Port of Charleston contribute to the area's economy?

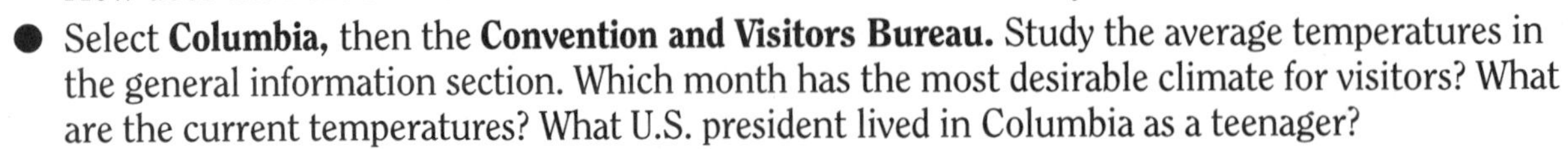

- Select **Columbia,** then the **Convention and Visitors Bureau.** Study the average temperatures in the general information section. Which month has the most desirable climate for visitors? What are the current temperatures? What U.S. president lived in Columbia as a teenager?
- ✎ Who was Celia Mann? What did she do? Where is her cottage located?

Web site: http://www.cityofclemson.org/

- Click on **Clemson Facts.** Where is the city located? What is its population? Go to **Clemson Gallery** for pictures of the area and downtown district.
- ✎ Research information about **Clemson University.** For what is the university best known? How does the university improve life in the city?

Cut along dashed line.

South Carolina

Tourist Attractions

Web site: http://www.sccsi.com/sc/

- Go to **Lowcountry and Resort Islands.** Read the **Welcome** and explain the topography of the Low Country. How does the climate affect the tourist industry? Which island is most popular with tourists? What sport is the most common?
- Go to **Historic Charleston.** What **Beaches and Parks** are near the city? Name five **museums** and explain what each one exhibits. How many **churches** are located in the city?
- Go to the **Olde English District.** Name the main cities in this region. What American Indians live in York County? What amusement park can be found here?
- ✎ Label the Low Country, Olde English District, Charleston, Clemson, and Columbia on a state map.

Web site: http://www.ego.net/us/sc/myr

- Learn about the area known as the **Grand Strand,** or take a **Quick Tour** of its beaches and major cities.
- ✎ Visit the links along the Grand Strand. Copy the map. Design icons to replace the names. Make a map key to identify the icons.
- ✎ What would you do for entertainment if the beaches were off-limits? Explain your choice(s).

South Carolina

Geography

Web site: http://s9000.furman.edu/hurricane/home3.html

- Read the information about **Hurricane Hugo.** What cities in South Carolina were damaged? What did the storm cost?
- ✎ On a map of the world, trace the path of Hurricane Hugo beginning off the coast of Africa.

Web site: http://www.intellicast.com/weather/seast/

- From this site you can check the four-day weather forecast for major cities in South Carolina. Read all the information. Check the **radar** maps for more specific information.
- ✎ Choose one day of the forecast and create a high and low temperature graph for all four cities.

Web site: http://www.lakesusa.com/areas/maps/scmap.htm

- What is the geographical location of South Carolina? What states and ocean border it?
- ✎ Use the **clickable state map** to locate lakes nearest to Charleston and Columbia.

Web site: http://www.tripsouth.com/maps/sc.htm

- What highway runs between Charleston and Columbia? How many miles would you travel between the two cities? How many miles is it from Charleston to Hilton Head traveling down the coast?

South Dakota

General Information

Web site: http://www.state.sd.us/state/capitol/capitol/tour/

- Enjoy a tour of the state capitol. Click on **Maps** to see a floor plan of the building. What is directly above the Governor's Portrait Gallery? On which floor is the First Lady's Collection?

Web site: http://www.state.sd.us/state/sdsym.htm

- Describe the colors and motif of the **state jewelry.** What is the state fossil? insect? What is the meaning of the **greeting,** "How Kola"?

Web site: http://www.state.sd.us/state/executive/tourism/sioux/

- Read the **Overview of the Great Sioux Nation.** Where did the nation originate? What is the correct name(s) for the nation of seven tribes?
- Go to **Milestones.** Make a chronological list of Sioux warriors and leaders.
- ✎ Search the Internet for additional information about the significance of the battle at Wounded Knee.

Search words: Black Hills, Mount Rushmore, Crazy Horse, Sioux Indians, Badlands National Park

South Dakota

Cities

Web sites: http://cityguide.lycos.com/rockymt/PierreSD.html
http://www.cam-walnet.com/~fpdc/

- Visitors to Fort Pierre will enjoy learning about the **history** of the area at the **Venendrye Museum.** Who were the Venendrye Brothers?

Web site: http://www.rapidcitycvb.com/

- In what county is **Rapid City?** What is its population? How many elementary schools are in the city school district? What air force base is nearby?
- Click on **Black Hills Attractions.** Visit several of the links.
- ✎ Write a profile of a typical tourist who would enjoy a vacation in the Rapid City area. What might be his or her background? interests?

Web site: http://www.siouxfalls.com/

- Read **Sioux Falls Statistics** and give five reasons why the city is a good place to live. What industries are major employers? How can you explain the low percentage of unemployment and lower than average cost of living?
- ✎ You may request a **visitor's packet,** or ask a question at the **information** link.

Cut along dashed line.

Cut along dashed line.

South Dakota

Tourist Attractions

Web site: http://www.state.sd.us/state/executive/tourism/rushmore/rushmore.html

- Whose faces are carved into **Mount Rushmore?** Who was Gutzon Borglum? Click on **Photo Album** for close-ups.
- ✎ What hazards did workers on the project face? How would such a project be different if it were attempted today?
- ✎ Give historical reasons to explain why each of the presidents was chosen to be included in the sculpture.

Web site: http://www.state.sd.us/state/executive/tourism/20reason/crzy_hrs.htm

- ✎ A monument to the Sioux warrior Crazy Horse is being carved into the Black Hills. Nearby is the Indian Museum of North America. Why is it appropriate for this type of monument to be located here?

Web site: http://www.usd.edu/smm

- **America's Shrine to Music Museum** is located at the University of South Dakota. You can see photos of many of its 750 instruments in a virtual tour of this site.
- ✎ How did the Industrial Revolution affect the production of musical instruments?

South Dakota

Geography

Web site: http://www.gorp.com/gorp/resource/US_National_Forest/sd_black.htm

- Learn about the **history** of the **Black Hills** from this site. What group originally owned the land? Explain how the discovery of gold led to loss of life and eventual occupation by the U.S. Army.
- Use the **Forest Overview Map** to locate three other state parks nearby. Name them. What highway runs northwest around the mountains?

Web site: http://www.state.sd.us/state/executive/tourism/sioux/artifact.htm

- Click on **Artifacts.** What types of artifacts can you view at Indian museums across the state? How is Indian art reflected in traditional clothing?
- Go to **Powwows.** What makes a powwow a good opportunity to learn about Indian culture? What must a spectator do to show respect for Indian customs?

Web site: http://www.state.sd.us/state/executive/tourism/sioux/snmap.htm

- From this clickable map you can read information about twleve **Points of Interest** of the Great Sioux Nation. Six reservations are indicated. Which ones are smallest? Which reservation crosses the northern border of the state?

Tennessee

General Information

Web site: http://www.state.tn.us/sos/history.htm

- Read about Tennessee **History** and summarize each section in two or three sentences. Be sure to look at the photo links.
- Who was **Andrew Johnson?** How was his career shaped by the assassination of Abraham Lincoln and the end of the Civil War?

Web site: http://www.state.tn.us/sos/symbols.htm

- This site has links to all of Tennessee's **symbols.**
- ✎ Choose five symbols to picture in a state book of facts. Add information from the Internet.

Web site: http://www.state.tn.us/education/webfacts.htm

- What U.S. presidents were born in Tennessee? When did Tennessee give women the right to vote? What is the highest point in the Smoky Mountains? What important scientific work took place at Oak Ridge during the 1940s?
- What is the origin of the name *Tennessee?* What is the population of the state? What are Tennessee's four major cities?

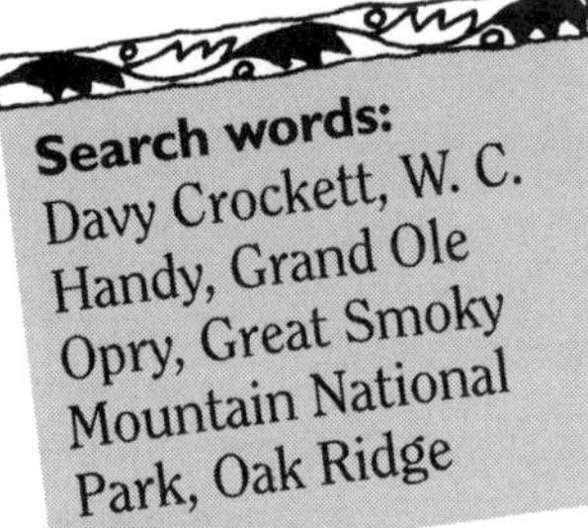

Search words: Davy Crockett, W. C. Handy, Grand Ole Opry, Great Smoky Mountain National Park, Oak Ridge

Tennessee

Cities

Web site: http://www.nashville.org/

- Who is the mayor of Nashville? Go to **Nashville History.** Choose ten people and write a one-sentence summary of the contribution each made to the city's history.
- Click on **weather.** What is the four-day forecast for Nashville?

Web site: http://www.downtownmemphis.com/

- Go to **History.** What is the Pinch? How did it get its name? Why is Beale Street famous?
- The **City Center Commission** is trying to attract conventions to downtown Memphis. Give three reasons why you might choose Memphis as a site for a business meeting.

Web site: http://www.chattanooga.gov/index.html

- Who is the **mayor** of Chattanooga? In what part of Tennessee is the city located?
- What is the four-day **weather forecast** for Chattanooga?
- ✎ Write a Tennessee weather forecast that could be read on the nightly news. Include Memphis, Nashville, and Chattanooga.

Cut along dashed line.

Tennessee

Tourist Attractions

Web site: http://www.grandoleopry.com/

- From this site you may read the **history** of the Grand Ole Opry or take a **backstage tour.**
- ✎ Choose one country artist. Search the Internet for biographical information and a list of recordings.

Web site: http://www.chattanooga.gov/index.html

- Go to **Links** and then **Ruby Falls.** Print a **Map** of the area and take the **Virtual Tour.** Read the **History** of the falls. Who was Leo Lambert? Who was Ruby?

Web site: http://www.elvis-presley.com/

- This is the official Web site for **Graceland,** home of Elvis Presley. Go to **Elvisology** and read his biography.
- ✎ Click on **Trivia** and make a list of statistics related to the success of his recordings and films.

Web sites: http://www.state.tn.us/tourdev/attract.html

- Click on **Oak Ridge** and read the list of **Attractions.** What is the importance of the **Historic Graphite Reactor?**
- Select the **Oak Ridge National Laboratory Tour.** Why was the laboratory established in 1942? What was the Manhattan Project?

Tennessee

Geography

Web site: http://www.tripinfo.com/maps/tn.htm

- Study this highway **map** of Tennessee. What interstate highway runs east-west across the state through three major cities? What two highways go south out of Nashville? What direction would you travel from Nashville to Chattanooga? How many miles would you travel?

Web site: http://www.chattanooga.gov/index.html

- Click on **Ruby Falls,** then **Cave Geology.** What caused the joints that formed Ruby Falls and Lookout Mountain Cavern? What factors affect the rate that cave deposits form?
- ✎ Make a chart showing diagrams and definitions for these cave formations: **stalactites, stalagmites, columns, drapes,** and **flowstone.**
- ✎ Click on **Fun Stuff For Kids** and make crystals in a dish.

Cut along dashed line.

Texas

General Information

Web site: http://www.texas-best.com/

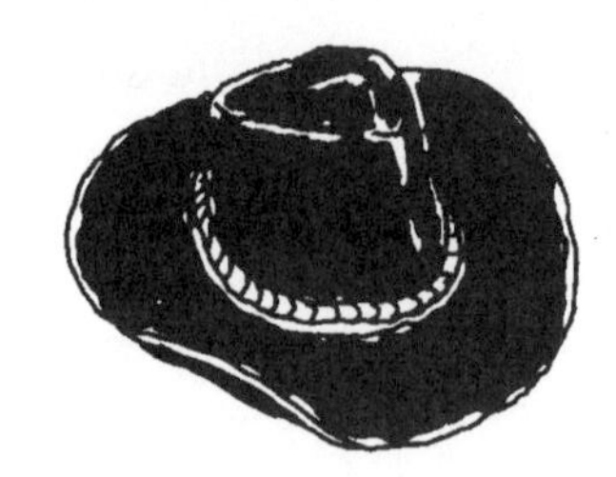

- Read the lists of **Texas Best Animals, Foods,** and **Plants.** Choose one category and combine the information into a *Texas Best* book.
- Go to **Cowboy Culture.** What is the job of a Texas cowboy? Explain the changes cowboys made in original Levi's® jeans and describe the trademark cowboy hat and boots.
- ✎ Make a dictionary of rodeo terms or horse breeds.

Web site: http://www.texas.gov/

- Click on **State Flag,** then look at the **Six National Flags of Texas.** Explain the historic significance of each one.
- ✎ Use art supplies to make a time line display of Texas flags.

Web site: http://www.governor.state.tx.us/

- Who is the governor of Texas? Select **Initiatives.** How does the governor hope to **"Turn around Texas"?** What is the Wrice Process?
- Why is the **Reading Initiative** important to the future of Texas?

Search words:
Lone Star State, cowboys, Rio Grande, Sam Houston, Alamo, Johnson Space Center, Colorado River Trail

Texas

Cities

Web site: http://www.ci.houston.tx.us/

- Select **City Government,** then **Office of the Mayor.** What are the duties of the mayor? What crimes are of greatest concern in Houston?
- ✎ Search the Internet and identify Sam Houston. Why is he important to the state's history?
- ✎ Draw a diagram to show the structure of Houston city government.

Web site: http://www.fortworth.com/Sundance/richards.html

- This very large site has information and photos of historical buildings in the Dallas-Fort Worth area. Look at the links for **Sundance Square.** What buildings make up these two blocks of Fort Worth's business district?

Web site: http://www.virtualelpaso.com/

- **Virtual El Paso** offers a wide range of information about this city near the Mexican border. Go to the **Site Map.** Read the weather forecast. Visit the **Historic Sites** and look at the **Beautiful Pictures.**
- ✎ What is the importance of **Tigua Tribe** and the Ysleta del Sur Pueblo? Where is their reservation?

Cut along dashed line.

Texas

Tourist Attractions

Web sites: http://www0.traveltex.com/TexasTours.html??mri0
http://www0.traveltex.com/PostcardsFromTexas.html?

- From this site you may choose any of six **tours of Texas.** Read the information and view the slides. Highlight the tour route on a map of the state.
- ✎ Choose and send a **postcard** that relates to your tour.

Web site: http://www.austin360.com/features/bats/batstop.htm

- This large site tells about Austin's major tourist attraction: the world's largest urban bat colony. Get **The Scoop on Bats** and view **Photos.** Click on **Interactive** to hear bat calls.
- ✎ Why do you believe people are fascinated by bats? If you were visiting Austin, would you take time to go to the bat bridge?
- ✎ Explain the ecological value of bats.

Web site: http://www.thedallaspage.com/

- Visit the Dallas-Fort Worth area from this site. What professional **sports** teams make Dallas their home? What is the purpose of the **Sixth Floor Museum?** Name two **colleges or universities** in the Dallas-Fort Worth area.

Texas

Geography

Web site: http://www.tpwd.state.tx.us/park/admin/historic.htm

- Click on **Archeological Sites.** What people lived at each of the sites? What artifacts have been uncovered? What can visitors do at each location?

Web site: http://www.tpwd.state.tx.us/nature/nature.htm

- Go to **Texas Critters** and read the **Wildlife Fact Sheets.**
- Make a list of **Threatened or Endangered Species** and explain what is being done to protect them.
- ✎ Print a map of Texas and draw ten animal or bird icons to show where they live in the state.

Web site: http://www.tripinfo.com/maps/tx.htm

- Print the state map. Label Mexico and the Gulf of Mexico. Circle the southernmost city in Texas. Estimate the number of miles between San Antonio and Austin; between San Antonio and Houston. What is the geographical location of Lubbock? El Paso? Texarkana?

Web site: http://www.lcra.org/lands/crt.html

- Study the map of the **Colorado River Trail.** It crosses 500 miles of Texas and connects eleven counties. Visit each of the county sites and make a list of featured attractions along the trail.
- ✎ Compare this map to a city map of Texas. What major cities are nearest the trail?

Utah

General Information

Web sites: **http://www.state.ut.us/html/utah_kids.htm**
http://ipl.org/youth/stateknow/ut1.html

- What do the beehive, bald eagle, and sego lily pictured on the **state flag** signify? What is the **State Animal? Bird? Dinosaur?**
- What is the population of Utah? What states border Utah?
- ✎ Choose a **Famous Utahn** and search the Internet for more information about his or her accomplishments.

Web site: **http://www.ce.ex.state.ut.us/history/uthist.htm**

- Go to **Mormon Settlement.** Who was Brigham Young? What caused the Mormons to move west? How did the group work together to settle new communities? How many immigrants had made the trip west by 1850? What did the Mormons name their state?
- Go to **Transition.** How did the population of Utah change between 1860 and 1890? In what year did Utah become a state?
- Go to **War and Depression.** Why was the depression especially difficult for people in Utah? How did the war provide jobs?

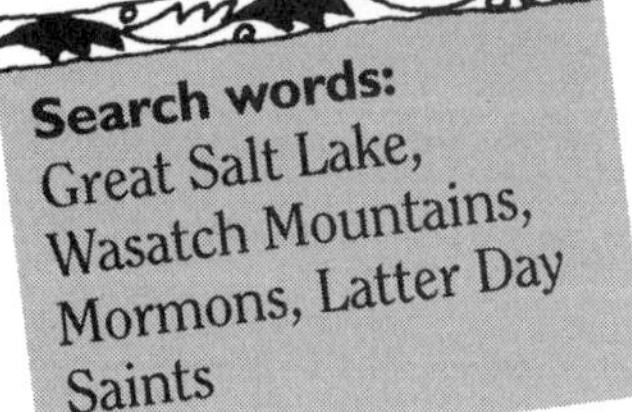

Search words:
Great Salt Lake, Wasatch Mountains, Mormons, Latter Day Saints

Cut along dashed line.

Utah

Cities

Web site: **http://www.ogden-ut.com/history.html**

- Read the **History** of the city. How did the Mormons contribute to early growth of **Ogden?** What was the effect of the Transcontinental Railroad? For whom was the city named?

Web site: **http://www.saltlake.org/**

- Click on **A Walking Tour of Salt Lake City.** What sites would you pass along the way? What buildings are located on **Historic Temple Square?**
- What is the city doing to prepare for the **2002 Olympics?**

Web site: **http://www.provo.org/**

- Click on **Economy.** Read the **Fact Sheet** and information about crime, employment, business, and education. Explain why **Provo** is one of America's most livable cities.

Web sites: **http://virtual.moab.ut.us/**
http://www.moabutah.com/

- Look at the photos of **Virtual Moab,** a city in the Canyonlands area. Go to **Parks, Monuments, and Museums** for more photos of the region. Describe the scenery in twenty-five words or less.
- What is the current news in Moab? Check the **Calendar.** What are three current local events?

Cut along dashed line.

Utah

Tourist Attractions

Web site: http://www.infowest.com/colorcountry/p3.html

- From these sites you can access many of Utah's state and national parks and outdoor attractions. What is the significance of the **Golden Spike National Historic Site?**
- Go to **Bryce Canyon National Park.** What American Indian legend explains the red rocks around the canyon? What colors can be seen in the limestone formations of the amphitheaters?
- ✎ Read the information about **Zion National Park.** Compare and contrast the Zion and Bryce Canyon parks on a Venn diagram.

Web site: http://www.dinopark.org/

- Click on **Photo Exhibits.** What dinosaurs are exhibited at **Ogden's Eccles Dinosaur Park?**
- Go to **Park Info.** A visitor can see more than a hundred full-sized dinosaurs in a park setting as they lived 200 million years ago. What special features make the dinosaurs seem lifelike?
- ✎ Which dinosaurs have been found in Utah? Choose one dinosaur and search the Internet for information about its behavior, diet, and habitat.

Utah

Geography

Web site: http://www.gorp.com/gorp/resource/US_nm/ut_natur.HTM

- Describe the location and scenery of **Natural Bridges National Monument.** What plant and wildlife can be seen in the park?
- Click on **Archaeology.** What remnants of ancient civilizations can be seen in the canyons?
- Read **Geology.** How did wind, water, and wildlife change the desert landscape? What kinds of plants survive in a **High Desert Environment?**
- Go to **How Old Is Old?** How does weather affect the erosion of sandstone? About how old are the bridges in the monument?
- ✎ What is the difference between a natural bridge and an arch?
- ✎ Search the Internet to learn about the Anasazi. Who were they? When and where did they live throughout the southwestern United States? Describe their homes.

Web site: http://wwwdutslc.wr.usgs.gov/greatsaltlake/saltlake.html

- This is the home page of the **Great Salt Lake.** Explain the lake's salt content and how it affects the presence of life there.
- How does the elevation of the lake affect its salinity and the rate of evaporation?
- ✎ Search the Internet for the name of at least one other terminal lake (a lake with no outlet) in the world. How is the environment changed by the salt content of the water?

Vermont

General Information

Web sites: **http://mole.uvm.edu/Vermont/vtgeog.html**
http://users.aol.com/glue5/flag.htm

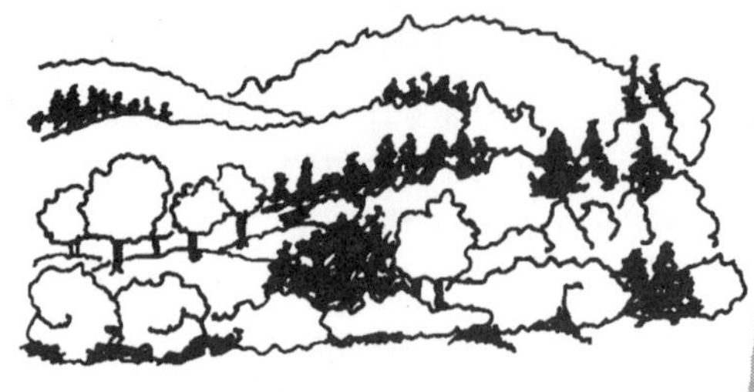

- Go to **History.** Who was the first European settler to arrive in Vermont? From what country did he come? What is the derivation of the word *Vermont?* Who were Ethan Allen and the Green Mountain Boys? In what year did Vermont become a state?
- Select **Official State Treasures.** Why is Vermont called the Green Mountain State? What is the state animal? Where are these horses bred today?
- ✎ Explain the meaning of the state motto "Freedom and Unity."

Web site: **http://www.virtualvermont.com/history/firsts.html**

- How has Vermont been a leader in education? Click on **People.** Name two U.S. presidents born in Vermont.
- ✎ Search the Internet for biographical information about John Deere, Norman Rockwell, or Stephen Douglas.
- ✎ Research and explain the function of a "normal" school.

Search words:
Mount Mansfield, Lake Champlain, Champ, Green Mountain Boys

Vermont

Cities

Web sites: **http://www.virtualvermont.com/towns/index.html**
http://users.aol.com/frotz/places.htm

- These sites offer an alphabetical listing of the cities, towns, and villages of Vermont. Choose any two of them to compare and contrast on a Venn diagram.

Web site: **http://www-burlingtonvt.together.com/**

- This is the home page for **Burlington.** What are the land area and population of the city? Who is the mayor? Where can you shop in Burlington? Who is Champ?
- ✎ Do research to understand the town meeting form of government common in many Vermont towns and cities.

Web site: **http://www.central-vt.com/towns/profile/ProMont.htm**

- Read the **actual population** (green) line on the graph. What does it tell you about population trends in Montpelier? Read **About Town.** What are Montpelier's major employers?
- Go to **History.** Who were the first people living in the Montpelier area? What artifacts have been found? Why is there very little evidence of their existence?

Cut along dashed line.

Cut along dashed line.

Vermont

Tourist Attractions

Web site: http://www.lilacinn.com/guide.htm

- This innkeeper's guide presents **Attractions in the Heart of Vermont.** Read about the stops on the Underground Railroad and the covered bridges.
- Have fun touring the **Teddy Bear Factory.** How did these bears get their name? Design a theme teddy bear for your state.
- ✎ Search the site of Brandon, Vermont, for more information about the Underground Railroad.
- ✎ Search the Internet for a photo of a Vermont covered bridge.

Web site: http://mole.uvm.edu/skivt-l/depths.html

- Read the graph showing **snow depths** atop Mount Mansfield.
- ✎ Use a calendar to fill in all the dates on the graph. Remember to advance the year. On what date was the snow its highest?

Web site: http://www.visitnewengland.com/vermont/north/places/inform.htm

- From this site you can access beaches, historic sites, attractions, and parks in the state. Select **Northern Vermont.** What are the current weather conditions?

Vermont

Geography

Web sites: http://mole.uvm.edu/Vermont/vtgeog.html
http://mole.uvm.edu/Vermont/GreenMount/
http://www.uvm.edu/whale/ZadockThompson.html

- Read **geographic facts** about Vermont. What are the highest and lowest points? What is the state capital? the largest city?
- Click on **glacial activity** and read about Charlotte, the Vermont Whale. What physical effects of glaciers can be seen in Vermont?
- ✎ Draw a sketch of the whale and explain where it was found and how it was preserved.

Web site: http://www.bluemap.com/maphome.htm

- From this site you can access **maps** of Champlain Valley, Greater Burlington, Downtown Burlington, and individual towns. From these maps you can zoom in for close-ups of the areas.
- ✎ Print any map. Write five questions about your map that can be answered with geographic directions. Pass your map and the questions to a classmate for answers.

Virginia

General Information

Web site: http://dit1.state.va.us/home/facts.html

- What is the geographical location of Virginia? How large is the state? What is the population density per square mile? What are the three physiographic areas of Virginia?
- Click on **Emblems of the Commonwealth.** What is the state dog? beverage? shell? flower? tree? Draw a picture of each one.

Web site: http://www.civilwar-va.com/

- Learn about **Virginia's Civil War Battlefields and Sites.** Read the **Historic Virginia** link. Choose three Civil War locations. Explain what took place at each site during the war and what can be seen there today.
- ✎ Draw a map of Virginia showing at least five Civil War battlefields.

Web site: http://www.baylink.org/Mattaponi/index.html

- Who was the first chief of the Mattaponi Indians? Why is the Mattaponi's story an important part of early American history? Where is their reservation? How large is it? How many tribal members live on the land today?

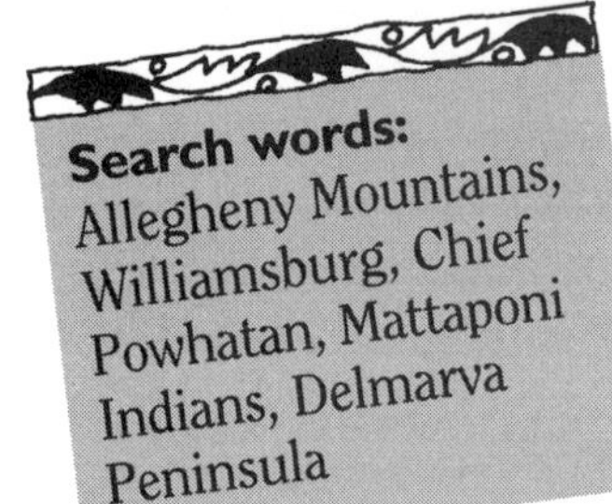

Virginia

Cities

Web site: http://www.ego.net/us/va/vb/index.htm

- Virginia Beach is the state's largest city. Go to **Climate and Geography.** Where is the city located? What are the average high and low temperatures for the three summer months? Read the **History** of the city. When and where did the first British settlers come ashore?
- ✎ Read the **Attractions, Festivals,** and **Cuisine** links. Explain how Virginia Beach attracts and accommodates tourists. Do you think it is a good tourist destination?

Web site: http://www.vgnet.com/norfolk/

- Read the **Introduction to Norfolk.** What bodies of water meet at the city's harbors? What naval base is located in Norfolk?
- ✎ Search for statistics about Norfolk's weather. Use the information to explain why the National Weather Service considers Norfolk to have the country's most desirable weather.
- ✎ Search the Internet for biographical information about Captain John Smith or Blackbeard.

Cut along dashed line.

Cut along dashed line.

Virginia

Tourist Attractions

Web site: **http://www.mountvernon.org/**

- See the home of George Washington on the **Mount Vernon Tour.** Go to **Archaeology.** What artifacts have been uncovered at the estate?

✎ Search the Internet for biographical information about George Washington. Compare the lives and interests of Washington and Thomas Jefferson.

Web site: **http://www.monticello.org/**

- Visit **Monticello,** the home of Jefferson. Read about **A Day In The Life,** or select **Matters of Fact** to understand more about plantation life.

✎ Choose one: Explain Thomas Jefferson's feelings about slavery.

Web site: **http://www.williamsburg.com/wol/tour/tour.html**

- What will a visitor experience at **Colonial Williamsburg?** What craftspeople are at work in the village? Go to **Historical Almanac** and read the biographies of famous people.

✎ Explain a typical day in the life of an eighteenth-century Williamsburg colonist.

- Go to nearby **Jamestown.** What hardships did early pioneers face? What can a visitor see in the English Gallery? Powhatan Indian Gallery? Triangular James Fort?

Virginia

Geography

Web sites: **http://www.chincoteague.org/**
http://www.atbeach.com/mdstpark/horses.html

- **Chincoteague Island** is a small fishing village and resort center. How many residents live there?

✎ Locate the island on a map. Describe its location and name two major mainland cities nearby.

- Read about the wild ponies of **Assateague Island.** View the photos. Where did the ponies come from? What do they eat? What is pony penning?

Web site: **http://www.luraycaverns.com/**

- Click on the candle icon to **Discover Luray Caverns.** Study the photos and read how the caverns were formed. Take the simple quiz along the way.
- Click on the compass for maps and directions to Luray Caverns.

✎ Use a highway map to plot a route from your hometown to Luray Caverns.

Web site: **http://www.naturalbridgeva.com/nathp.htm**

- Go to **General Information.** Describe the size of Natural Bridge. Go to **Map** and explain the roads and directions you would travel from Natural Bridge to Monticello and Luray Caverns.

✎ Natural Bridge is one of the Seven Natural Wonders of the World. Search the Internet and name the other six.

Washington

General Information

Web site: http://www.wa.gov/features/kids.htm

- Select **Washington State Symbols.** What is the state nickname? What is unusual about the capitol building? Explain how the goldfinch became the state bird. Explain the symbolic colors of the state tartan.
- ✎ Learn to sing the state folk song "Roll on Columbia, Roll on."

Web site: http://www.tourism.wa.gov/general/students/facts.htm

- At this site you can access **general information** about Washington's economy, resources, government, and environment. What are the most important agricultural crops? mineral resources? What are Washington's three largest cities?
- Who is the **Current Governor?**
- Go to **History.** What Native Americans lived in the region that is now Washington? Why did Europeans first explore the territory? In what year was Washington separated from the Oregon Territory?
- ✎ Explain how the Transcontinental Railroad influenced settlement of the Northwest.

Search words:
Gingko Petrified Forest, Olympic Peninsula, Dungeness River, Cascade Mountains, Puget Sound, Lake Chelan, Grand Coulee Dam

Cut along dashed line.

Cities

Web site: http://www.olympiaonline.com/intro.html

- **Olympia** is the capital of Washington State. Read **History.** When did Captain Peter Puget discover the sound? Who was living there at that time? What did they call the area?
- Name the early American settlers who laid out the city.
- ✎ Suppose you had settled a new territory. Design a street plan showing city buildings, residential neighborhoods, and shopping districts.

Web site: http://seattle.yahoo.com/

- What **Professional Sports Teams** make their home in Seattle? Choose one team and search the Internet to find its record for the most recent (or current) season.
- Read about Seattle's **Communities.** Study the links for five of the communities. Explain the demographics and attractions in each one.

Web site: http://www.ci.tacoma.wa.us/

- Read **About Tacoma.** Describe the **Climate and Environment.** What are the major employers? What **Fun Things** can you see and do in Tacoma? What events take place at the Tacoma Dome?
- Click on **Port of Tacoma.** How does the port link Tacoma with the rest of the world? What products are exported and imported there?

Cut along dashed line.

Washington

Tourist Attractions

Web site: http://www.olympiaonline.com/intro.html

- Go to **Government,** then **Tourism.** Read the list of **Fun Things to See and Do** in Olympia. Would you enjoy **Whale Watching?** Explain.
- Go to **Weather.** Describe the typical weather conditions. What kinds of clothing should a tourist pack for a spring vacation?
- ✎ Make a list of **Fun Things to See and Do** in your hometown. Use the telephone directory and include addresses.

Web site: http://www.spaceneedle.com/

- Why was the **Space Needle** built? How is it like the Eiffel Tower in Paris? Who were the artist and the architect who designed and built the structure? Go to **Fun Facts** and read **The Colors.** Make an accurately colored drawing of the Space Needle.
- ✎ Design a souvenir or postcard that might be sold at the Space Needle.
- ✎ What is Sneedle? Explain how and why it was built.

Web site: http://www.gorp.com/gorp/resource/US_National_Park/wa_mount.HTM

- About 2 million people a year visit **Mount Ranier National Park.** What **recreational activities** are possible there? How much of the park is covered with glaciers? Describe a volcano that is dormant but not extinct.

Washington

Geography

Web site: http://vulcan.wr.usgs.gov/home.html

- Use **Visit a Volcano** for links with descriptions and maps of the **Cascade Range Volcanoes** in Washington State. Choose one site to describe in detail and locate it on a map.

Web site: http://www.tourism.wa.gov/general/students/facts.htm

- Go to **Physical Geography.** What is the highest point in the state? Where is the Cascade Range? What is the topography of the area? Where is the Columbia Plateau? What is the topography of that region?

Web site: http://www.wa.gov/puget_sound/aboutps/whatis.html

- Describe **The Terrain** of Puget Sound.
- Read **Habitats and Environmental Values.** What is an estuary? Explain the shoreline environment and describe the wildlife it supports. Locate the San Juan Islands on a map.
- ✎ What is the current status of **Water Quality** in Puget Sound? Look at the map. Click on one of the five regions for specific information.

Cut along dashed line.

Washington, D.C.

General Information

Web site: http://www.nps.gov/parklists/dc.html

- Here is a complete list of **National Park Service Units** in the capital. The map shows their locations. Which former presidents have memorials in the nation's capital?
- Click on the **National Mall.** Go to the **National Mall Home Page** link. Who designed the mall? How many acres does it cover? What events are held there each year? What museums face the mall?

Web site: http://www.his.com/~matson/hoods.htm

- How are the towns and neighborhoods around Washington influenced by the history of the capital? What nearby states are considered to have Washington, D.C., neighborhoods?

✎ Use a map of the region to locate these cities.

Web site: http://www.ci.washington.dc.us/

- Click on **About Washington, D.C.** Read the **Chronology.** On what date did George Washington select the site? From what states was land taken?
- Who is the **mayor** of Washington, D.C.?

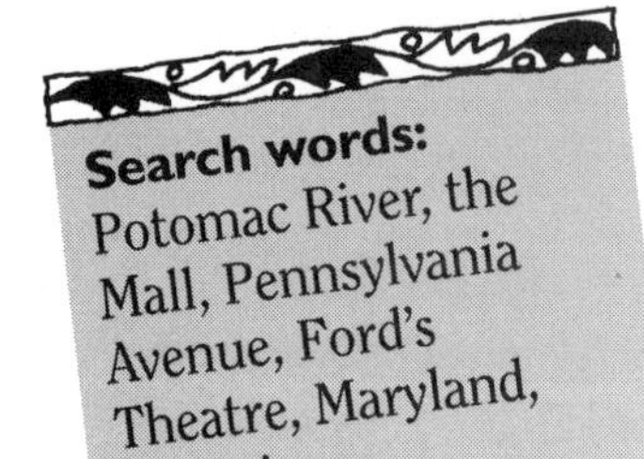

Search words:
Potomac River, the Mall, Pennsylvania Avenue, Ford's Theatre, Maryland, Virginia

Washington, D.C.

Historic Sites, Landmarks, and Museums

Web site: http://www.si.edu/

- Go to **Events and Activities.** What should your family know before visiting the Smithsonian?
- Go to **Resources and Tours.** Select **A Kid's Guide to the Smithsonian.** Tour the museums of Natural History, American History, and Air and Space. What is the address of each museum?

✎ Make a chart or a web to illustrate ten items exhibited in each museum.

Web site: http://www.nps.gov/wamo/index2.htm

- Describe the **Washington Monument.** What is an obelisk? a colonnade? Name the architect who designed the monument.

Web site: http://www.nps.gov/linc/index2.htm

- Explain the symbolism of the **Lincoln Memorial.** Of what is it made? Who was the architect? What artist made the statue of Lincoln?

Web site: http://www.nps.gov/vive/index2.htm

- Where is the **Vietnam Veterans Memorial** located? Who designed it? When was it built? How are the names organized on the memorial?

Cut along dashed line.

Washington, D.C.

Government Buildings

Web site: http://www2.whitehouse.gov/WH/Welcome.html

- Enjoy a **White House Tour** and read the **history** of the building. Click on the cutaway diagram. What will you see in the Vermeil Room? the East Room? the Diplomatic Reception Room?
- ✎ Make a chart listing all fourteen rooms and areas open to visitors. Indicate the original use and contents of each room and choose one important feature of the room to explain in detail.
- Go to **Art in the White House,** then **Selected Works from the Collection.** List one piece from each category, the artist, and the benefactor.
- ✎ Send an e-mail greeting to the president and his family.

Web site: http://www.aoc.gov/

- This site has considerable information about the history and construction of the **United States Capitol** building and grounds.
- Who designed the **dome** in the capitol? How much did it cost? Who was president at the time of its construction? Look at the photos of the dome.
- ✎ Who was the landscape planner for the grounds? What problems did he face when he began the project? How have the grounds changed in the twentieth century?

Washington, D.C.

Geography

Web site: http://www.sbpm.gwu.edu/Map/

- Use this large clickable map for links to all of Washington's government buildings and historic landmarks.
- ✎ Locate the White House. With that starting point, what direction would you travel to reach the U.S. Capitol? the Vietnam Veterans Memorial? the JFK Center for the Performing Arts?

Web site: http://sc94.ameslab.gov/TOUR/tour.html

- Click on the map for an interactive map of the Washingtone DC area. Select any neighborhood to view a close-up map and information about the community.
- ✎ How is Adams-Morgan/DuPont Circle different from Georgetown? In which community would you prefer to live? Explain.

Web site: http://www.his.com/~matson/canal.htm

- Read **Background Information** about the **Chesapeake and Ohio Canal National Historic Park.** Where does the canal begin and end? What was the original purpose for the canal?
- ✎ How has the canal area been damaged by flooding? What is being done to restore the area for outdoor enthusiasts? Use the links to learn the details.

West Virginia

General Information

Web site: http://www.state.wv.us/

- What are the state bird, animal, fish, and tree?
- ✎ Draw an outdoor scene including these symbols. Add your idea for an appropriate state insect. Explain your choice.
- ✎ Why would a miner and a farmer be pictured on the state seal?

Web site: http://members.aol.com/jeff560/wv-fam.html

- From this site you can learn about **famous people from West Virginia.** Use the alphabetical listing to identify an athlete, a musician, a politician, a scientist, and a television personality.

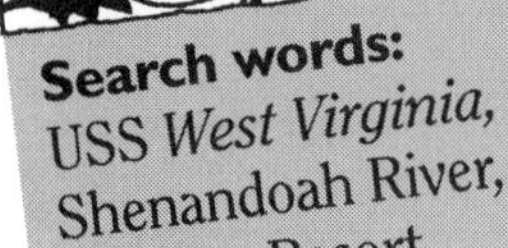

Search words:
USS *West Virginia*, Shenandoah River, Oglebay Resort, Appalachian Mountains

Web site: http://www.wvcivilwar.com/

- Read about West Virginia's role in the Civil War. Which side did most West Virginians choose to support? View the large group of photos and illustrations.
- Explain how **West Virginia** achieved **statehood.** How was the problem of slavery overcome? Who was U.S. president at the time?
- ✎ Who was Ann Jarvis? Design an appropriate greeting card to explain Ms. Jarvis's contribution to American culture.

Cut along dashed line.

West Virginia

Cities

Web site: http://wvweb.com/www/cities/cities.html

- Click on **Charleston,** capital of West Virginia. Go to **History.** How did West Virginia become a state? What other city has been the state capital? Describe the capitol building. What is unique about its dome?
- Select **The City** and read the **Welcome.** What is the population of the city? What highways meet in Charleston? Go to **Landmarks.** What historic buildings line Capitol Street? What is the significance of Shrewsbury Street?

Web site: http://www.olcg.com/wv/wheeling.html

- Select **Wheeling-The City.** Read the information and explain how Wheeling's location has influenced growth and change over its history. What **Attractions** draw visitors to this river city?

Web sites: http://wvweb.com/www/cabell_huntington_cvb.html
http://cityguide-att.lycos.com/midatlantic/HuntingtonWV.html

- Describe the location of **Huntington.** What industries are important to the city? What university is located in the city? Read the **History** of the area, then go to **Points of Interest.** What is unusual about the East End Bridge?
- ✎ Name three organizations that might schedule a convention in Huntington.

Cut along dashed line.

West Virginia

Tourist Attractions

Web site: http://www.wvculture.org/

- The **State Museum** helps preserve West Virginia's art and culture. Where is it located? Who was the **Hatlady?** What did she donate? Read about the permanent **Exhibits** on the USS *West Virginia* and glassmaking.

Web site: http://wvweb.com/www/travel_recreation/cave.html

- West Virginia offers many **caving** opportunities. Read these links to understand what tourists see in their underground adventures. What preparation(s) would you need to make to enjoy exploring a cave?
- Read about the **Pocahontas Mine.** When was the mine opened? How long did it produce coal? Who worked in the mine?
- ✎ How has mining and energy use changed over time?

Web site: http://www.nps.gov/hafe/hf_visit.htm

- Explain the historical importance of **Harpers Ferry.** Where is it located? What **notable people** made history at Harpers Ferry? Who was John Brown?
- Go to **Maps and Information.** What two rivers meet in the park? What three states include part of the park? Take the **Virtual Tour** to understand the important story of this town.

West Virginia

Geography

Web site: http://wvweb.com/www/seneca_caverns/web.html

- **Seneca Caverns** are the largest in the state. Read information about the **geological history** and underground **formations.** How do scientists know the caves are still growing and changing?
- ✎ Make a cave dictionary with these terms: flowstone, calcite, travertine, stalagmites, stalactites, calcium carbonate, and crystals.

Web site: http://www.nps.gov.neri/welcome.htm

- Enjoy a hike through the **New River Gorge** with a park ranger. Go to **Natural Resources** and make a list of birds, animals, and plants in the park. Highlight the endangered species.
- Click on **Geology.** Where does the water originate? How has the movement of the water created the gorge? How long do geologists believe the gorge has been forming?
- ✎ Read the information and trace the water's 2,000-mile journey on a map from its headsprings in North Carolina to the Gulf of Mexico.

Web site: http://www.try-rivers.com/map.html

- Print the map of West Virginia and surrounding states. How many miles is it from Charleston to Washington, D.C.? to Columbus, Ohio? to Paintsville, Kentucky? Which direction(s) would you travel to each city from Charleston?

Cut along dashed line.

Wisconsin

General Information

Web sites: **http://www.ipl.org/youth/stateknow/wi1.html**
http://www.ipl.org/youth/stateknow/wi2.html

- What is the origin of the state's name? What does it mean? What are the major industries in Wisconsin? What states border Wisconsin? Who are some famous Wisconsin natives? Why is milk the state beverage?
- ✎ What is a badger? Why is Wisconsin called the Badger State?

Web site: **http://www.state.wi.us/agencies/tourism/guide/faq00.htm**

- At this site of **Frequently Asked Questions** you can learn the highest elevations in the state; the names of lakes, forests, and parks; and rainfall and snowfall averages.

Web site: **http://www.wisconline.com/**

- Select a subject from the pull-down menu and click to see what events are taking place this month in Wisconsin. You can also access a weather report and a clickable county map.

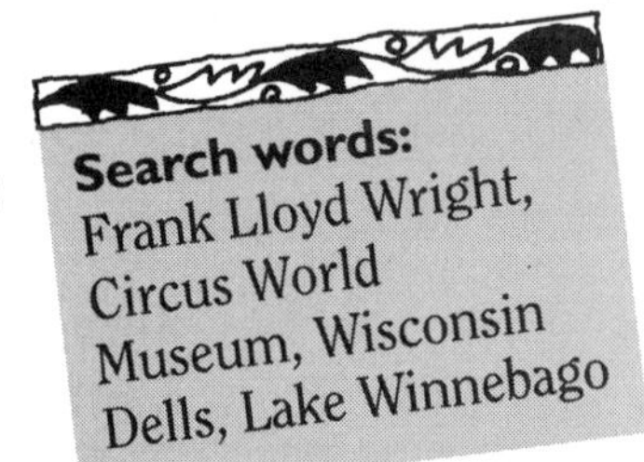

Search words:
Frank Lloyd Wright, Circus World Museum, Wisconsin Dells, Lake Winnebago

Wisconsin

Cities

Web sites: **http://www.visitmadison.com/**
http://www.ci.madison.wi.us/

- Visit **Madison Wisconsin's Convention and Visitors Bureau** and click on **Greater Madison at a Glance.** What is the combined area of the city and its lakes? What university is located in the city? What **Recent Awards** has Madison won?
- What is the current weather forecast? What are the average temperatures for each season?
- ✎ Explain this statement: Madison is the only North American city built on an isthmus. Draw a map to illustrate your explanation.

Web site: **http://www.milwaukee.org/**

- Go to **Visitor Information,** then **Community Background.** Who is the mayor of Milwaukee? Go to **City History.** What festivals reveal the city's ethnic heritage? What is the origin of the city's name?
- Go to **Milwaukee Today.** What is the city's geographical location? How important is tourism to the city's economy?

Web site: **http://www.msoe.edu/~reyer/mke/**

- Tour **Historic Milwaukee Architecture** from this site. Choose your two favorite structures from each time period and describe them in detail.

Wisconsin

Tourist Attractions

Web site: http://tourism.state.wi.us/agencies/tourism/

- Click on **Wisconsin Best Bets,** then **Top Ten Destinations.** Choose one destination that interests you and search the Internet for more detailed information.

Web site: http://flw.badgernet.com:2080//

- **Frank Lloyd Wright** was born in Wisconsin. Use the site index to view photos of some of his buildings. Click on the **Taliesin** link and describe Wright's architectural style.
- ✎ What training is necessary to become a professional architect?

Web site: http://www.mpm.edu/

- Tour the **exhibits** of **Milwaukee's Public Museum.** Create a poster or diorama of a scene from one of the exhibits. Go to **Other Permanent Exhibits** for a look at cultures around the world.
- ✎ Suppose you worked at this museum. Choose an exhibit that interests you and prepare a brief speech welcoming visitors to it.

Web site: http://www.dells.com/

- The city of **Wisconsin Dells** is a popular family vacation spot. Go to **History** and read the American Indian legend about the forming of the Dells. Describe the natural beauty of the area.

Cut along dashed line.

Geography

Web site: http://www.newnorth.net/glitc/

- Read about each of the **Native American Tribes** in Wisconsin. Make a chart with information about each tribe's festivals, entertainment, and tourist attractions.
- ✎ Label a state map with the names of the reservations.

Web site: http://www.dnr.state.wi.us/fh/fish/

- Read about the species of **Wisconsin Fish.** Choose five species and chart their spawning habits, where they are found, and the best way to catch them.

Web site: http://www.nass.usda.gov/wi/

- Wisconsin is an **agricultural state.** Study the chart to determine what percentage of the total land area is farmland. How many acres of farmland are cropland? woodland? pasture?

Web site: http://www.wisconline.com/maps/counties.html

- Use this site for **maps** of Wisconsin's counties and cities. Select a county in the north, south, east, and west of the state. Is there an airport? a lake or river? a major highway? How many miles wide is the county at its widest point?

Wyoming

General Information

Web site: http://www.state.wy.us/state/welcome.html

- Click on a county for a **Virtual Tour** of Wyoming. Visit three counties. Summarize their demographics and environment. Locate Yellowstone National Park and Cheyenne, the state capital.
- ✎ List five ways Wyoming differs from your home state.
- Go to **Wyoming Info,** then **General Facts.** Explain the culture of the Plains Indians who lived in the state. What American Indians live on the Wind River Reservation?
- ✎ Explain in detail why Wyoming is nicknamed the Equality State.

Web site: http://commerce.state.wy.us/tourism/collage.htm

- At this site you can view the **state symbols** as well as photos of its most spectacular natural landmarks.
- ✎ Briefly explain how Yellowstone National Park was formed.
- ✎ What do the people pictured on the state seal symbolize?

Search words: Yellowstone National Park, John Colter, Kit Carson, Fort Laramie, Grand Teton Mountains, Devils Tower, Wyoming State Museum

Wyoming

Cities

Web site: http://www.laramie.org/

- Click on the **City of Laramie** and read about the city's history. For whom was the city named? Briefly describe the development of the early community. What kinds of businesses operate in **Laramie Today?** View **Photos of Laramie.**

Web sites: http://www.lcc.whecn.edu/scc/Cheyenne/
http://www.cheyenneweb.com/

- Read the **Ten Most Wanted List** in the **Places To Go** link. Which activities are related to cowboy culture?
- **Cheyenne Web** brings you the news of the area. What are today's local headlines?
- ✎ Explain the work of a cowboy, such as jobs on the range.

Web site: http://www.casperwyoming.org/city/

- Read the links for **Casper, Wyoming.** Go to **Casper Town Square.** What is the population of the city? What business, recreation, and educational opportunities are available in the city?
- ✎ Write five questions you would like answered at the Visitors Center.

Cut along dashed line.

Cut along dashed line.

Wyoming

Tourist Attractions

Web site: **http://www.jackson-hole.com/**

- Jackson Hole is a popular tourist destination. Watch the **Slide Show.** What recreational activities are available in the area? Read the history of Grand Teton **National Park.** How did John D. Rockefeller help build this park and preserve its environment?

Web sites: **http://www.yellowstone-natl-park.com/**
http://www.muscanet.com/~lross/w9410814.htm

- Enjoy the links of the **Total Yellowstone Page.** Read the rules for camping and exploring the **Backcountry.** When were the most recent **wolf and bear sightings?**
- ✎ What must a student do to become a **Junior Ranger?**
- View a **Gallery** of photos of **Old Faithful** geyser in **Yellowstone National Park.**

Web site: **http://www.gorp.com/gorp/location/wy/wy_nw.htm**

- Click on the **Best of the West** icon. Select the map of regional **Wyoming.** Use the clickable map to access links for national parks, monuments, historical sites, forests, and mountain ranges in Wyoming. Through what cities does Highway 80 pass? What cities are located near the border of Montana? South Dakota? Idaho?
- What national forest is nearest Casper? What national park is nearest Jackson?

Wyoming

Geography

Web site: **http://www.web-net.com/jonesy/geysers.htm**

- This page has information and photos of the nine **geyser basins** in Yellowstone. Use the links to read **How Geysers Work** and the **Glossary of Terms.**
- ✎ Draw a map of the geysers in Yellowstone.
- ✎ What geologic conditions help to form geysers?
- ✎ Search the Internet to define the caldera rim.

Web site: **http://www.jackson-hole.com/parks/geology.shtml**

- Read the information about the **Geology of the Teton Mountains.** How old is the range? What is a diabase? Explain the effects of the Ice Age and glaciation on the Jackson Hole area.

Web site: **http://www.nps.gov/yel/**

- What geological features does Yellowstone National Park encompass? Go to the **Fact Sheet.** How much of the park is in Wyoming? What other states have some park acreage? How many active geysers are in the park? Look at the **Map.** Describe the location of Old Faithful and Yellowstone Lake.
- ✎ Design a series of six U.S. postage stamps commemorating Yellowstone, the world's first national park. Feature a different attraction on each one.

Glossary

Bookmarks: A feature that saves and organizes addresses so that you can find them quickly at a later time (sometimes called Favorites).

Browser: A term that identifies the kinds of software you use to access the HTML pages.

Electronic Mail (E-Mail): A means of sending messages from one Internet user to another.

Frequently Asked Questions (FAQs): Text containing answers to questions commonly asked about a specific topic.

Gopher: A text-only information system.

Graphics Interchange Format (GIF): An image file format.

Home Page: The opening page of a Web site. It contains links to additional and related information.

Hypertext: The words on the screen that are a different color from the rest. They may be underlined. If you place the cursor on these words or phrases, a pointed finger will appear. Click the mouse and download an image, sound, or another Web page.

Hypertext Markup Language (HTML): Language used to create Web pages.

Hypertext Transfer Protocol (HTTP): The way your web browser speaks to the Web server program on the World Wide Web.

Icon: A small picture or symbol that represents something. Point your cursor at the icon, click, and start the application.

Internet: A matrix of networks that connects computers around the world.

Modem: Equipment that connects a computer to a telephone line.

Mouse: A small device that connects to the computer and directs the movement of the cursor around the screen.

Online: Connected to the Internet, either sending or receiving information.

Search Engine: A program that uses key words to search all Web pages and sort out sites having the information you have requested.

Uniform Resource Locator (URL): The address of a Web site.

Web Site: The location on the World Wide Web where you can find specific information.